The Countryside: Planning and Change

The Resource Management Series
Editors: Richard Munton and Judith Rees

The Countryside: Planning and Change

Mark Blacksell
Senior Lecturer in Geography,
University of Exeter

Andrew W. Gilg
Lecturer in Geography,
University of Exeter

London
GEORGE ALLEN & UNWIN
Boston   Sydney

First published in 1981

GEORGE ALLEN & UNWIN LTD
40 Museum Street, London WC1A 1LU

British Library Cataloguing in Publication Data

Blacksell, Mark
    The countryside. – (The resource management
series; No. 2).
    1. Regional planning – Great Britain
    I. Title   II. Gilg, Andrew W   III. Series
    711′.3′0941 HT395.G7   80–41220

    ISBN 0–04–711008–2

Typeset in 10 on 12 point Times by Typesetters (Birmingham) Limited
and printed in Great Britain
by Hollen Street Press Limited, at Slough, Berkshire

Foreword

The Resource Management Series reflects the resurgence of interest in resource analysis that has occurred over the past twenty years in both the natural and the social sciences. This interest mirrors wide public concern about declining environmental standards, man's detrimental impact on the ecosystem, the spatial and temporal allocation of resources, and the capacity of the Earth to sustain further growth in population and economic activity.

Academic research should play a crucial role in policy formulation if informed decisions are to be made about resource use or about the nature and pace of technical and economic change. The need to assess the impact of technological developments on the environment is widely recognised; this cannot be done in physical terms alone but must involve social science research into the economic, social and political implications. Failing this, society may persist in trading off environmental gains for more easily definable economic advantages, an option which is particularly tempting in times of slow economic growth, high rates of inflation and rising unemployment. Furthermore, a planned approach to resource use makes the study of policy − its formulation and implementation − imperative; and this requires a sound understanding of the options available, the legal and administrative contexts, the decision-making behaviour of planners and managers, and the day-to-day realities of the decision-maker's environment.

Cost−benefit analysis, landscape evaluation, environmental impact assessment, systems modelling and computer simulation techniques have all advanced significantly in recent years as tools of resource analysis. Although none of these are without their deficiencies, they have undoubtedly improved our understanding of the effects of resource utilisation decisions and of the complex interrelationships that exist within and between the physical and economic systems. Moreover, their use has clearly indicated that effective inquiry in the resources field cannot be confined to any one discipline.

The Series has been planned as an interdisciplinary vehicle for major contributions from scholars and practitioners with a wide variety of academic backgrounds. The Series is unequivocally directed towards policy formulation and management in the real world, and it will not include contributions which merely describe an economic or physical resource system, or those which are entirely theoretical in nature. However, the subject area is defined widely to include the management of all natural resources, renewable and non-renewable, whether managed by private enterprise or public-sector agencies.

It is hoped that the books appearing in this Series will command the serious attention of all students, scientists, planners, resource managers and concerned laymen with an interest in understanding man—environment interactions and in improving our resource decisions. Each book draws on substantial research or practical management experience and all reflect the individual views and styles of the authors. The editors and publishers hope that the Series will not only encourage further research but will also play an important role in disseminating the results.

In their book, Mark Blacksell and Andrew Gilg examine the effects of post-war planning measures on the use and the management of rural land in Britain. These cover a wide range of issues including rural settlement structure, service provision, landscape protection and wildlife conservation. But as the authors point out, what is conspicuous by its absence is any agreement over priorities. As a result most initiatives have proved inadequate and powerful interest groups have been able to promote and maintain partisan policies. Consensus has rarely been more than skin deep and it is with the major reason for this, the fragmentation of policy making, that this book is primarily concerned.

The authors explain the countryside's complex administrative and legislative structures and analyse the economic and technological forces responsible for today's agricultural and afforested landscapes. Evidence from their own studies in Devon is combined with those of others working elsewhere in Britain. These all suggest that, contrary to public opinion, the rate of landscape change in the urban fringe or on the moorland edge is no greater than in the countryside as a whole. In the more specific context of settlement policies and the effectiveness of development control as a planning tool, Blacksell and Gilg conclude that although scattered developments in the countryside have been minimised, more general urbanisation trends have been scarcely modified. One reason for this is the lack of control held by planning authorities over the provision of many public services and utilities, such as education and water supply.

What makes this book such an important contribution to the Series, and to the present debate on how to plan the countryside, is its presentation of so much original material. Its publication is also timely, as the Government, under the aegis of the Countryside Review Committee, is currently reappraising its approach to rural land management. The Committee favours the continued search for consensus, as reflected in the use of management agreements between private landowners and public bodies, but the authors, on the basis of their wide research experience, suggest that the management agreement approach, whilst representing a step forward, is unlikely to be successful in the absence of greater co-operation between sectional interests, and change in this regard can only originate in Whitehall.

RICHARD MUNTON and JUDITH REES
April 1980

Preface

This book assesses the impact of the policies and plans that have proliferated during the past 30 years to guide the process of change in the rural landscape of Great Britain. Many public and private agencies are deeply involved with the future of the countryside, yet despite widespread interest and concern, co-ordinated decision-making has been notably lacking. As the pressures on rural land have multiplied and intensified, so the lack of concerted policies has become ever more apparent, along with inevitable clashes of interest.

The research that formed the starting point for the study was a detailed examination of land use and settlement change in the County of Devon, supported by the Social Science Research Council between 1972 and 1977. It has allowed the findings of other people in different parts of the country to be placed in perspective. Without SSRC support the book would never have been written and we are very grateful for the opportunity it gave us.

Over such a long period many people, especially our colleagues in the Geography Department at the University of Exeter, have contributed to our efforts in innumerable different ways and we should like to say a general 'thank you' to them all. Some, however, must be singled out: Avril Sinclair, our research assistant between 1975 and 1977; Dave Leat, whose PhD research was a considerable inspiration to us both; and Phipps Turnbull, Chief Planning Officer at Devon County Council, whose generosity in making available primary data was a crucial factor in the success of the research. We have also been helped greatly by our editors, Richard Munton and Judith Rees, who performed the unenviable task of coaxing a readable end-product from our first scribblings with much patience and skill. Roger Jones, the Science Sponsor at George Allen & Unwin, has also been a constant source of help and encouragement. In the last few months Maureen Cook has typed the manuscript, converting our untidy offering into a professional finished product. Finally, we thank our families who have lived with both the research and the book for so many years and received so little credit for making it all possible.

MARK BLACKSELL and ANDREW GILG
Exeter, December 1979

Contents

Contents

List of tables

List of illustrations

1 *Introduction*

A developing concern for the countryside

Change in the countryside is not a new phenomenon, but a widespread general concern about the scale and nature of the changes has been a relatively recent post-war development. When Wordsworth protested about the introduction of the railway to the Lake District in the mid-nineteenth century he reflected the feelings of a tiny group of people. Even at the end of the century, the newly founded National Trust was the work of a few dedicated pioneers, who would have found it incredible that the membership could grow as large as 25 000 by 1950, let alone the 540 000 members that the Trust had in 1975.

Nineteenth-century concern for the countryside was restricted mainly to an intellectual and generally affluent minority and it was not until the 1920s and 1930s that a mass 'amenity movement' was born. This movement had its roots in two areas: first, the newly affluent and mobile middle classes of home counties suburbia, and second the dedicated Sunday ramblers of the nothern industrial towns. The hallmarks of the new class were mobility and a determination to enjoy the combined benefits of both the urban and rural environments. Most members of the new class were employed in towns and cities, but they romanticised about the countryside from which their forefathers had come only two or three generations previously and, to satisfy their yearnings, they tried to recreate a 'rustic idyll' in the suburbs and in those villages within commuting distance of the major centres of employment. The net result was a breakdown of the division between town and country, and the gradual erosion of the power base of rural landowners, as urban populations extended their influence into the countryside. The growing strength of the mobile middle class also generated conflicts about the proper use of the countryside. The pre-eminence of agriculture and forestry were increasingly questioned, as middle class groups were organised to lobby the nation and Parliament. Before the new movement could really become established however the Second World War intervened, diverting the mass energies in other directions. It also gave time for reflection and stock-taking and was an essential precursor to the burst of legislative activity in the late 1940s.

In the last 30 years the division between town and country has been further weakened with an ever growing proportion of the population living in rural or semi-rural areas, almost all of them exercising their right to visit the countryside during their much more numerous leisure hours. The countryside has become common property and concern for both it and the rural way of life in general has been popularised in a way with which most people can identify. Unfortunately, the image of the countryside generally portrayed by the media of a gentle, unhurried way of life, free of most of the pressures of contemporary society, never really existed. The contrast between this image and the reality now being revealed to holiday makers and day trippers, goes a long way towards explaining the upsurge of disquiet about changes in the countryside. This was admirably captured in the 1978 'Strutt report' *Agriculture and the countryside*, 'There is evident concern about the harmful effects of many current farm practices upon both the landscape and nature conservation . . . a widespread feeling that agriculture can no longer be accounted the prime architect of conservation nor farmers accepted as the natural custodians of the countryside . . . It had now to be recognised by all those involved in the countryside that the countryside is an area of legitimate concern for a whole variety of interests' (Advisory Council 1978, p. 27).

The drift away from the cities and a wholly urban life style that began in the 1930s developed into a flood after the war, as modern telecommunications and high speed travel allowed people to live and work outside the urban centres and to enjoy the fruits of urban civilisation in rural surroundings. As a result there are now large numbers of people living in the countryside, who care passionately about its future preservation, but who are not part of it in so far as they are employed and earn their livings wholly in the towns and cities. Indeed, the new mobility and affluence mean that there are more people than ever before concerning themselves with the future of rural areas, but ironically this has happened at a time when a combination of technological, economic and political factors are precipitating more changes in the countryside than at any time in the past 300 years.

Aims of the book

This book sets out to examine in detail the nature and dynamics of the new patterns of agricultural land use and settlement referred to at the end of the previous section. It should be made clear from the outset that a fundamental distinction is drawn between the forces controlling land use changes in agriculture and forestry, and those determining housing development in the countryside, for each is subject to different sets of powers and, for the most part, governed by separate agencies. For the purposes of this book the two are termed the 'open countryside', where farmland, forestry and moorland dominate the scene, and the 'developed countryside', where villages, small towns and all the paraphernalia of the built environment hold sway. In both cases, however, ownership, or more precisely development rights in the widest sense, are fundamental in determining how the land use will change and the extent to which demands for change will be satisfied.

In the 'open countryside' agriculture remains the major user of land and decisions about future change lie largely in the hands of farmers and landowners. They are, however, influenced by a wide variety of outside agencies, the most important of which are the Ministry of Agriculture and the European Commission in Brussels, whose systems of price regulation, grants, subsidies and advice guide the pace and direction of agricultural change. Nevertheless, in spite of these outside influences farmers remain freer than many other sections of society to alter the appearance of their land. They can put up buildings that would not be granted planning permission for non-agricultural purposes; they can grub up hedges and drain marshes with almost total impunity, indeed it is usual for such activities actually to attract government aid. The net result of these and other freedoms has been to arouse much disquiet, especially in the non-agricultural agencies of the government, such as the Department of the Environment (the DOE), the Nature Conservancy Council and the Countryside Commission, all of which have a duty to safeguard specific aspects of the rural environment. Forestry, the other major land user is more directly controlled through the policies of the Forestry Commission and, since the mid-1960s, these have become more sympathetic to wider landscape and social issues. The futures of both agriculture and forestry are dominated by two issues: the conflicting responsibilities and advice given by the different government agencies and how to manage better the nature and rate of land use change (Ch. 4).

Demands for change in the 'developed countryside' are much more complex, because of the much larger number of interests and individuals involved. Most important is the construction industry in the form of either private builders seeking to develop land for profit, or local authority housing committees trying to shorten housing waiting lists. In both cases they first need to find land and persuade the owner to part with it at an acceptable price. They then require planning permission, which will in turn hang on the specific provisions of the statutory plans for the area and the degree to which the site is serviced by public utilities, such as electricity, gas, water, sewerage and communications. Some places are better provided than others and the legal plan (either the old style Development Plan or the newly introduced Structure Plan) will take this into account, as well as other considerations such as the desirability of safeguarding beautiful landscape and good agricultural land, and the need to provide for economic growth. Once developers have overcome these initial hurdles and the houses are actually built buyers have to be found, either people employed locally, or those willing to commute, or wishing to retire to the country, or even the growing numbers seeking a second, holiday home. The complex interplay between builders, planners, landowners, house buyers and tenants is examined in detail in Chapters 5 and 6.

The differences between the processes of change in the 'open' and 'developed' countryside have created two distinct types of rural management:

(a) Resource planning. This encompasses the development of the countryside's primary industries, subject to the needs of landscape preservation, nature conservation and recreation.
(b) Development planning. Here the purpose is to exercise some measure of public control over the development of rural land for such things as urban growth, water supply and mineral extraction.

Very few studies have attempted to look at both these aspects of rural land management together although a good deal of work has been conducted into individual problems like agricultural land use change, the losses of agricultural land to urban development, and the conversion of open moorland to enclosed farmland. The aim of this book is to explain not only the processes, but also the mechanisms for managing countryside change and then to proceed to a detailed examination of the relationship between them in a number of case studies, with particular reference to work carried out by the authors in Devon during the 1970s.

Devon as an area for case studies

The sheer size of the County of Devon and the variety of its landscape, agriculture and settlement pattern make it a very useful laboratory for studying the processes of land use change. Within its boundary it has the whole of the Dartmoor National Park and a part of the Exmoor National Park, three Areas of Outstanding Natural Beauty and a fourth soon to be designated, four stretches of Heritage Coast, an almost continuous Coastal Preservation Policy Area along both the north and south coasts, nine Areas of Great Landscape Value, five National Nature Reserves, two Forest Nature Reserves, one Local Nature Reserve, and nineteen of the nation's eighty-nine Sites of Special Scientific Interest.

The main industry is agriculture and ranges from arable farming on the Grades 1 and 2 land along the Exe valley in the south-east, to dairying and livestock rearing in mid- and south Devon, and to upland grazing on the bleak moorlands of Dartmoor and Exmoor. Many conflicts arise, because most of the most valuable agricultural land lies close to the major cities of Exeter, Plymouth and Torbay and along the axis of the M5 motorway where the pressures for other forms of development are greatest. Elsewhere much of the poorest land that for decades has been ignored for intensive cultivation, is now capable of economic reclamation and the controversy over the loss of moorland on Exmoor is symptomatic of changes taking place throughout the uplands. New agricultural techniques have also dramatically reduced the number of farming jobs in Devon as elsewhere in the country, even though there were over 750 applications between 1974 and 1977 to build new agricultural dwellings, a paradox that is examined further in Chapter 6.

The settlement pattern is also a microcosm of that in the rest of Britain with the four main features of rural life all represented. First, Devon has one of the most prosperous and fastest growing urban areas in the country in the region between Exeter and Torbay, with the rapid spread of development leading to losses of good agricultural land and the change of agricultural villages into dormitory estates. Secondly, there is an area of static or declining population in mid- and west Devon, where the problems of providing jobs and services for a dwindling population are all too similar to those that beset Wales, northern England and most of Scotland. Thirdly, sections of the southern coast between Exmouth and Budleigh Salterton and between Dartmouth and Plymouth are under severe pressure to accommodate second and retirement homes and other forms of tourist development. Fourthly, the more accessible parts of the protected landscapes described above are the scene of a classic conflict, with on the one hand developers trying to capitalise on their protected status, while on the other preservation interests seek to protect them from all assaults. Grafted on to this complex picture is an annual influx of over 3·5 million tourists and the recreational wants of the local population, which result, in Dartmoor alone, in 8 million day visits a year.

Devon is also the scene of a number of more specific land use planning controversies: the use by the military of about 14 000 ha of the Dartmoor National Park; the conversion of moorland to farmland on Exmoor; the mining of china clay on the south-western edge of the Dartmoor National Park; the construction of reservoirs at Wimbleball on Exmoor and Bickleigh near Plymouth; and road improvement schemes that have taken and threaten to take either Grade 1 agricultural land, or land in the National Parks.

All these pressures have forced the county council to conclude that 'The rural environment has always been subject to change; what is different now is the pace and scale of change. Pressures are coming from more and more sources and *they have intensified*. The scale of individual projects, whether for a new road, reservoir, quarry, or industrial site is bigger than before. Pace of change and scale of change combine to threaten the rural environment in a way it has never been threatened before. Many of these pressures arise from the need to maintain and improve the standard of living of communities' (Devon County Council 1977, p. 116). Given such an awareness on the part of the public authority, it is not surprising that there has been a readiness to try to plan and control the changes, which makes for a stimulating environment in which to study the problems and possibilities for rural land management.

Introduction

The nature of the countryside

In spite of the major incursions of urban life styles into rural areas, the British countryside still remains fundamentally different from urban areas in both appearance and intensity of use. Based on these differences, this book defines countryside as those areas, 'which show unmistakable signs of being dominated by extensive uses of land, either at the present time or in the immediate past' (Wibberley 1972). This includes not only agriculture and forestry, but also other uses requiring large tracts of rural land, such as

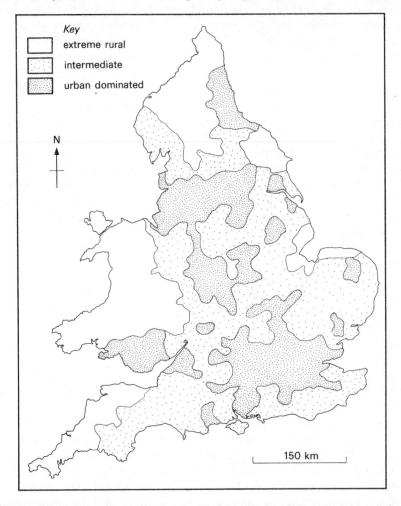

Figure 1.1 An index of rurality for England and Wales. Sixteen variables derived from the population census are analysed by principal components analysis to produce an index of rurality. The diagram shown here has been simplified from the original work based on local authority boundaries. (Cloke, P. 1977. An index of rurality for England and Wales. *Regional Studies* **11**, 31–46.)

water supply, recreation, nature conservation, defence training, and mining, nuclear and oil-based industries.

The countryside can also be subdivided according to population pressure and the ensuing degree of urban dominance as shown in Figure 1.1. In this analysis sixteen variables thought to reflect differences between rural and urban environments are grouped together using cluster analysis (Cloke 1977a). Cloke's original map has been simplified by reducing the number of area types from five to three, so as to avoid the almost inevitable confusion arising from the use of the very small pre-1974 local authority boundaries. The result is three types of rural area, each roughly equal in area. First, the extreme rural areas (usually remote and in the uplands) where urban influences are slight, except for the temporary impact of summer tourism and weekend recreation. Secondly, the intermediate rural areas (usually farmed lowland) where there is a balance between rural and urban interests. Thirdly, the areas where urban influences dominate (usually in the urban fringe and the extended commuter fields of the larger cities) where the superficial appearance of the countryside is unaltered, but where the communities are dominated socially and economically by urban values. For the purposes of this book, a fourth area has been created by further subdividing the remote areas into the hills and the uplands respectively, because of the different planning and management problems posed by either predominantly enclosed or unenclosed landscapes.

Cherry has said that 'The countryside has a popular image of unchanging tranquility. This is almost part of our difficulty, because we have to learn to appreciate that the countryside is in fact changing rapidly; the myths, impressions and beliefs which have been transmitted to our present generation have now to be systematically explored by careful study in order to expose a contemporary reality' (Cherry 1976, p. 2).

This book attempts to do just that and draws on many studies of land use and housing in the countryside to provide an assessment of the real nature and scope of change in rural areas.

2   The administrative context

There are three main facets to the government's approach to countryside change: the imposition of negative controls, the offering of positive financial inducements, and the proffering of advice. The most popular is the last, persuasive advice, since it is usually the least costly and, potentially, the most effective in a free society where controls and regulations ought to be the weapons of last resort. All three are now considered, both in terms of the intrinsic nature of the organisations charged with administering rural change and in terms of the areas over which the respective organisations exercise jurisdiction.

Ministry of Agriculture, Fisheries and Food (MAFF)

The need to administer and oversee the development of agriculture was first recognised in the 1930s, but full realisation of how necessary an expansionist agricultural policy was to Britain did not dawn until the naval blockade of the Second World War interrupted the flow of cheap food that had freely entered the country since the repeal of the Corn Laws in the mid-nineteenth century (Tracy 1976).

The bulk of present government policy is founded therefore on the Agriculture Act 1947, although modified by the 1957 and 1963 Acts and, since Britain joined the European Community in 1973, by the provisions of the Common Agricultural Policy (CAP). Agricultural policies are worked out centrally by MAFF, but are administered locally by the officers of the Agricultural Development and Advisory Service. There are also about 30 additional boards, tribunals, committees, commissions and councils, supervising the activities of different sectors of the industry (Civil Service Department 1976).

The formulation and administration of agricultural policy has become a good deal more complex since Britain joined the European Community, largely because of the fundamental differences between agriculture on the European mainland and in the UK. The original members of the Community are more or less self-sufficient in temperate foods and even net exporters of some of them, which makes these countries largely independent of world agricultural markets. The structure of farming is also quite different, with holdings being on average smaller, and so consequently there are greater numbers of farmers and farmworkers. The large labour force makes the agricultural lobby much stronger in Europe than in the UK, and the farming unions have been able to exert much more influence on policy than the National Farmers' Union in Britain (Mackintosh 1970). Finally, the idiosyncracies of the CAP system of support mean that both open and guaranteed market prices are usually higher, and often much higher on the continent.

To help smooth out the differences, British entry into the Community was phased between 1 February 1973 and 31 December 1977, but even so there are still considerable discrepancies between the two systems in both overall approach and detail. Movements in exchange rates, mainly between the US dollar, the German Mark and the pound sterling, have confounded all attempts so far at achieving a common price structure. In 1978 German agriculture prices were on average 7% above the norm, while British and Italian prices were respectively 23% and 16% below (Swinbank 1978), although in the course of 1979 British prices have moved closer to the European average. In general, however, individual governments have been so successful in maintaining their own policies that in 1978 seven price zones were in operation, rather than a single unified structure.

The administration of the system is extremely complex and annual negotiations about the level of price support, the exchange rates to be used (Monetary Compensation Amounts) and the levies to be imposed on food imports, are lengthy and each year take at least 3 months to complete. The resulting policy, derived from what is in essence a very inefficient political compromise, is administered in Britain by the Intervention Board for Agricultural Produce and elsewhere in the Community by FEOGA, the European Agricultural Guidance and Guarantee Fund (European Communities 1978). For analytical purposes, the net result is that the true effect of agricultural policies on changes in UK agriculture structure are less easily distinguished than they were in the years between 1947 and 1973 when the industry was solely a national concern.

In spite of Community membership the Agriculture Act 1947 remains an important policy guideline, for it lays down the government's official objectives for the industry, 'A stable and efficient agricultural industry capable of producing such part of the nation's food and other agricultural produce as in the nation's interest it is desirable to produce in the United Kingdom, and of producing it at minimum prices consistent with proper remuneration and living conditions for farmers and workers in agriculture and an adequate return on capital invested in the industry' (Section 1). The main weapon for achieving these objectives is the control or guidance of prices so that a balance between imported and home produced food prices is struck, which suits the needs of the balance of payments, government expenditure and incomes in agriculture, as well as providing prices acceptable to the domestic consumer and the food processing industry (Bowman 1976, Houston 1975).

The negative controls used by the government consist mainly of systems of quotas and maximum prices. For example, quotas are imposed on the acreage that individual farmers may allocate to hops, sugar beet and potatoes. In the case of milk, both a quota and a maximum price operate in a system administered by the Milk Marketing Board which buys up nearly all the country's milk supplies. Each year a level of output is agreed and excess production over this is sold to manufacture cheese, butter or skimmed milk. As milk and its by-products, including beef (about 75% of home produced beef comes from dairy herds), are so central to modern British agriculture, the influence of the price controls is felt throughout the entire industry. They make for stability and for the certainty of a reasonable and regular income from dairying and are a central factor in explaining the popularity of this type of husbandry. Fortunately, from the point of view of landscape preservation, dairying demands fewer landscape changes than most other enterprises, but new techniques like zero grazing and the production of artificial fodder may in the future reduce the need for grassland husbandry based on traditional fields.

Positive inducements to agricultural production are of two main types, the price support system that helps ensure a reasonable return on production, and the direct payment of production grants and subsidies to aid and to influence to some extent the nature of the production process itself. Contemporary price support is very different from what was originally envisaged in 1947. There have been three main phases:

(a) 1947–73 The deficiency payment or guaranteed price system,
(b) 1973–77 The transition from deficiency payments to CAP intervention payments,
(c) 1978–   Intervention payments. Minor deficiency payments for those few products outside CAP.

The deficiency payments system was so useful and effective that it was only abandoned with great reluctance on entry to the Community and the British government would still like to see the CAP move towards its introduction throughout Europe. Its great advantage was the direct relationship with market forces, which meant that food prices tended to fall to the advantage of the consumer without the farmer suffering from uneconomic returns. This also prevented the creation of unsaleable surpluses and excessive support bills for the government. The system worked by fixing a guaranteed price and, if the open market price fell below it, the government paid the difference in the form of a deficiency payment. As the guaranteed price could only be varied marginally from year to year, the system introduced stability and confidence into the farming industry. There is little doubt that it encouraged the investment needed to modernise farming systems, especially in arable farming, and was therefore instrumental in producing rapid landscape changes in the eastern half of the country.

Deficiency payments were agreed each year at an Annual Review, when the Minister of Agriculture and representatives of the farming industry met to review production trends and market requirements, world market prospects, the cost of subsidies, income trends in agriculture, efficiency and changes in production costs (Cmnd 7058 1978). The procedure is still followed but executive power to implement most change has largely passed

Table 2.1 Public expenditure on agricultural policy (£ million). (Cmnd 7058 1978. *Annual review of agriculture 1978*. London: HMSO.)

	1973–4	1975–6	1977–8 (forecast)
price guarantees	113·50	10·0	3·8
production grants and subsidies	64·3	87·1	63·4
support for capital and other improvement	87·9	73·7	77·6
support for agriculture in special areas	40·7	30·1	58·9
market regulations under CAP	85·6	310·7	203·9
less receipts from FEOGA	76·7	264·0	173·80

to the European Commission, through the system of intervention prices, which were gradually introduced between 1973 and 1977, as shown in Table 2.1 (Central Office of Information 1977). This has led to an increase in expenditure because the CAP does not have the same checks and balances as are found in the deficiency payment system. The central feature of intervention prices is that once open-market prices fall below the intervention price (fixed just below a target price) the Community has to buy all produce offered to it at the intervention price. It not only provides a floor to the market, but also a guaranteed sale and this has encouraged overproduction and created large surpluses, especially in milk and milk products. The problem is compounded by the high levels set for intervention prices as a result of pressure from the powerful farming lobby, and by the exclusion through import levies and quotas of the much cheaper supplies of food available on the world market. However, British farmers did not enjoy the full benefit of these high prices till 1979, because between 1973 and 1977 the European Community agricultural currency (the green pound) was devalued less than the pound sterling on the open market.

The complexity of the new system makes it difficult to evaluate, particularly since it has been the subject of constant government manipulation both in London and Brussels (House of Lords 1977, Marsh 1977). Certainly it has not led to the widespread increases in the area devoted to arable farming that was first thought.

In the meantime the British government continues to try to act as if the UK were not a party to the CAP and, in a 1975 white paper, proposed a further long-term expansion of agricultural production at a rate of 2·5% per annum and, somewhat ironically, emphasised the need for increased milk production (Cmnd 6020 1975). Although at odds with the CAP, which aims to reduce the overall Community milk surplus, the objective was logical in purely national terms, since Britain is far from self-sufficient in either butter or cheese. These policies were broadly confirmed by a further white paper in 1979, which although it dropped precise targets, hoped for up to a 10 to 20% increase in production in the next 5 years (Cmnd 7458 1979). However, the only real certainty in the confusion surrounding agricultural policies in the 1970s is that farmers can no longer rely on government for firm and direct guidance, because its hands are tied by its European commitments, and this seems bound to slow down investment in new machinery and production systems.

The one area where MAFF has remained firmly in control is in the payment of production grants and subsidies. In landscape terms the most notorious of these grants have been those for hedge removal, for ploughing up moor and grassland, and for land drainage. In combination, they have not only hastened the trend towards field enlargement and hedge removal in existing arable areas, but also allowed arable systems to encroach on land previously too poor for cultivation due to limitations of slope, drainage or climate. Currently, these more specific payments are being phased out in favour of more general grants such as the Farm and Horticulture Development Scheme, the Farm Capital Grant Scheme (that allows grants of up to 70% in the less favoured upland areas of the country), the Farm Structure Scheme and the Hill Livestock Compensatory Allowance. Since many of these initiatives are intended to improve the profitability of marginal land, they conflict with other policies to conserve landscape and wildlife, particularly in National Parks and other protected landscapes.

In common with all the other members of the Community, Britain's agricultural policies have recently been bogged down in the annual argument over short-term price levels, but it is long-term objectives that are more important when it comes to planning future farm structure. Structural reform is vital and has mainly been achieved through the efforts of the Agricultural Development Advisory Service backed up by judicious grants (Helme 1975, Winifrith 1974). Undoubtedly the scientific knowledge and enthusiasm of the 5800 staff have encouraged farmers to modernise their holdings and inaugurate landscape change. Indeed, their role is so central to the implementation of agricultural policy and thus to countryside change, that it has been proposed that their remit be widened to provide a general rural affairs advisory service (Advisory Council 1978). Such a change would reduce the conflict their advice has sometimes led to in areas like Exmoor and would be a logical extension of the European Community advisory service that already provides finance for the employment of rural socio-economic advisers in the countryside.

The Agricultural Development and Advisory Service is complemented by a very wide range of other advisory and scientific organisations. Marketing boards help with distribution, sales and quality control; financial credit is provided by the Agricultural Mortgage Corporation; and pests and diseases are controlled and, hopefully, eradicated by MAFF itself (MAFF 1976a). Basic and applied research to the tune of £65 million in 1976–7 is carried on at experimental farms and the results subsequently released to farmers, help to make British agriculture the most scientific in western Europe.

Overall government policy towards agriculture has invariably encouraged expansion and greater intensification of land use, which have led directly to the landscape changes discussed in Chapter 4. There is little doubt that British agriculture would be far less advanced and progressive were it not for government involvement, but success itself has created problems, notably in relation to the rate, scale and impact of landscape and wildlife change caused by the new husbandry. The myth that modern agriculture is the best insurance policy for the continued existence of the traditional countryside is now scarcely credible, especially when the large reduction in the workforce is also taken into account. There are, however, signs that both government and the farmers are coming to realise that increased production has been bought at too high a price and discussions are beginning to take place to find how modern farming practices and conservation can co-exist (Ch. 7).

Forestry Commission

The Forestry Commission has administered the development of forestry since its foundation in 1919. Initially, the Commission was charged solely with the creation of a strategic timber reserve, but since the 1950s its role has been progressively widened and now includes such things as employment, amenity, recreation and conservation. So successful has it been, that the Commission is now frequently cited as a model of how to integrate and manage conflicting land use demands in the countryside within a single agency.

The Forestry Commission is still dependent on financial aid from the government and its work is nominally overseen by MAFF and the Scottish and Welsh Offices, but as its early forests come to saleable maturity, it is becoming increasingly self-financing. There are two main sections: a Forest Authority, charged with promoting the interests of forestry, the development of afforestation and the production and supply of timber; and a Forest Enterprise, which actually plants, manages and markets the products of the Commission's own forests. By 1977 these covered 0·84 million ha or 42% of all woodland (DOE 1978c). Since 1972 the planting programme has been reduced to 22 000 ha a year and, as the replanting of felled forests rather than the expansion into new areas becomes the norm, the rate of growth in the forest estate will inevitably slow down.

Under the Forestry Act 1951 the Commission also exercises major control over private forestry through the system of felling licences. Most trees over 10 cm in diameter cannot be felled without such a licence. Normally a licence will impose a replanting condition, although this does not prevent deciduous trees being replaced with more profitable coniferous plantations.

Another control over felling, and one more applicable to small woodlands and individual trees, is the Tree Preservation Order, which first appeared in the Town and Country Planning Act 1947. This enables local planning authorities to restrict the owner of a tree or group of trees from felling or otherwise damaging them. The power has been widely used by the authorities when sanctioning developments like new village estates, so as to give them a sense of scale and continuity. As with felling licences, Tree Preservation Orders may be rescinded and a replanting condition substituted.

Such measures are, however, no more than a stop gap and far more can be gained by positive encouragement to plant new trees, for trees like any other plant will eventually die. Accordingly, the Forestry Act 1947 made provision for Dedication Agreements, whereby private owners could dedicate their forests to a system of management supervised by the Forestry Commission, in return for grant aid. About 600 000 ha of private woodland out of a total of just over 1 million ha were dedicated in 1977. In spite of the fact that dedication agreement grants had been largely instrumental in producing high rates of planting, over 10 000 ha per annum in the 1950s and 1960s, they were temporarily abandoned in 1972 when a review of forest policy concluded that neither private nor public forestry could be justified on purely economic grounds (MAFF 1972a, HM Treasury 1972). There was an immediate outcry, which led to the scheme being revised and reintroduced as the Basis III system in 1977 (Forestry Commission 1977a). Basis III is confined to areas above 10 ha only, but provides grant aid of £100 per hectare for conifers and £225 per hectare for broadleaves. In return, owners have to agree to manage their trees according to sound forestry practice, to integrate the forest with agriculture, to safeguard the environment and to provide appropriate opportunities for public recreation.

Woods under 10 ha are covered by a Small Woods Scheme, designed to promote and preserve their fabric and, since most of them are broadleaved, they are vitally important to the landscape and ecology of lowland farming areas. Below 0·25 ha, Countryside Commission grants for copses and other small planting schemes complement grants for shelter belts made by the Ministry of Agriculture (Countryside Commission 1975).

A variety of tax concessions provide further incentives for afforestation, the aim being to make some kinds of forestry more attractive than agriculture, although the replacement of Death Duties by Capital Transfer Tax in 1975 removed many of the advantages. The new tax extended taxation to lifetime transfers of capital and, of most importance to forestry, to the underlying value of the land. Even though trees are not taxed until felled and despite a few concessions since 1975, the new system reduced the rate of planting by 1976–7 to only 62% of the 1970–4 average (Campbell 1975, Forestry Commission 1977b). Further concessions led to a recovery by the late 1970s.

Finally, the Forestry Commission exerts enormous influence over forestry through its research programmes and pioneering management skills. Research is beginning to transform timber production into a highly efficient operation, with improvements similar to those achieved in arable farming. Indeed, scientific progress threatened for a time to destroy the traditional multiple use of forests as wildlife, amenity and recreational resources. However, the wider responsibilities now conferred on the Commission have aroused a deeper and wider concern for the countryside and have led to broadly based management plans, which aim to integrate both public and private forestry more closely into the landscape as a whole. As a result it is now accepted that the commercially less favoured land ought to be used for recreation, or for the planting of broadleaves so as to encourage a greater variety of wildlife. Nevertheless, the end result is still not always appreciated by the public and afforestation in National Parks, for example, is severely curtailed, even though forests can absorb greater numbers of visitors and harbour a larger if different fauna than the moorland they often replace.

Department of the Environment

Within the DOE responsibility for most countryside matters lies with the Planning, Sport and Countryside Directorate. The Directorate advises on policy for the conservation and enhancement of the natural beauty and amenity of the countryside and of its fauna, flora and geological and physiographical features, provision for public access and open air recreation, National Parks, caravans and camping, trees and hedgerows, public rights of way and commons. A number of essentially independent but related bodies also advise on these matters and provide advice to the DOE, notably the Countryside Commission and Nature Conservancy Council, and Table 2.2 provides an outline of their expenditure. In addition

Table 2.2 Official expenditure on recreation and conservation (£'000). (Countryside Commission and Nature Conservancy Council 1978, 1977. *Annual reports 1976–77, HC 273 and HC 53.* London: HMSO.)

	Countryside Commission		Nature Conservancy Council	
	1975–6	*1976–7*	*1975–6*	*1976–7*
salaries and administration	648	650	3006	3608
research and experiments	189	269	780	882
information and visitor services*	472	616	201	568
grants (or loans)	1552	1817	33	42
national parks supplementary grant†	2700	3400	230	205

*For NCC, Capital expenditure and acquisition of nature reserves.
†For NCC, Maintenance of properties and nature reserves.

there are over a dozen agencies providing advice to the DOE and the subsequent fragmentation of effort and policy has been criticised as being a wasteful use of resources (House of Commons Paper 1976). In Scotland and Wales the pattern is more centralised within the Scottish Development Department and the Welsh Office.

Countryside Commission

The Countryside Commission for England and Wales replaced and added to the functions of the National Parks Commission under the provisions of the Countryside Act 1968. It is a statutory body charged with keeping under review all matters relating to the provision and improvement of facilities for the enjoyment of the countryside, the conservation of its natural beauty and amenity, and the need to secure public access. It has not always been clear whether the main emphasis in the Commission ought to be recreation or conservation and at times it has found it difficult to reconcile its multi-purpose duties (Cripps 1980). A Commission for Scotland was set up in 1967 under the provisions of the Countryside Commission (Scotland) Act 1967. The work of both Commissions can be grouped under 8 headings:

(a) Overseeing the workings of the land use planning system,
(b) Designating and advising on policies for National Parks and Areas of Outstanding Natural Beauty (not Scotland),
(c) Assisting local authorities and private owners by providing grant aid for the provision of recreational facilities or the improvement of the landscape,
(d) Promoting information and interpretation facilities,
(e) Providing long-distance routes,
(f) Advancing knowledge about countryside recreation and conservation by commissioning research and providing in-service training,
(g) Studying and advising on countryside issues of national importance,
(h) Liaison with other government agencies and with voluntary movements (Countryside Commission 1978a).

Nature Conservancy Council

The Nature Conservancy was brought into being by Royal Charter in 1949 and, after a number of changes in the 1960s, was reconstituted as the Nature Conservancy Council under the Nature Conservancy Council Act 1973. Like the Countryside Commission it is a statutory grant-aided body, but unlike the Commission it covers Scotland as well as England and Wales. The Council's main functions are the establishment, maintenance and management of nature reserves, the provision of advice to the government on the development and implementation of policies for, or affecting, nature conservation, the provision of advice and the general dissemination of information about nature conservation matters, and the commissioning and support of relevant research (DOE 1974c, 1977d). However, most research was given to the newly formed Institute of Terrestrial Ecology when the Nature Conservancy was restructured in 1973, leaving the Council with nature reserves and the general provision of advice as its main duties. The two roles are directly related, for current Council advice is that unless the machinery for a national rural land use strategy is implemented soon the present policy of isolated conservation in separate protected reserves will fail (Nature Conservancy Council 1977a).

National Trust and Amenity Societies

The National Trust founded in 1895 has now grown to become Britain's largest private landowner with 220 000 ha under its control as well as over 200 historic buildings and nearly 700 km of coastline. Its 540 000 members demonstrate the depth of concern felt by the British public over undue change in the countryside, and the National Trust has led the way in promoting positive conservation of the landscape by developing new techniques of landscape and visitor management. The Trust enjoys special powers, which allow it to declare its land inalienable. It is a unique right, which theoretically prevents its sale, even to the government, though sadly the government itself has overridden the right in two cases in recent years.

Other groups and organisations have to exert their influence more indirectly, for they do not usually own much land in their own right. They may be broadly divided into two. First, there are the professional organisations, like the Royal Town Planning Institute and the Royal Institution of Chartered Surveyors, that set professional standards, represent the interests of their members, and where necessary exert pressure on government for legislative change. Then there are the pressure groups that exist solely to represent the interests of their members, unencumbered by professional ethics. The Town and Country Planning Association, the Council(s) for the Protection of Rural England and Rural Wales, the Ramblers' Association, the Royal Society for the Protection of Birds and many other groups are dedicated only to the furtherance of their own ideals and these may well change with time. Collectively, such organisations exert considerable pressure on central and local government, encouraging both to take their particular interests into account when making land use decisions.

All pressure groups have been assisted by the Countryside Act 1968, the Forestry Act 1967 and the Water Act 1973 that for the first time required public agencies to have due regard for the needs of conservation and recreation when coming to a decision about future land use. Unfortunately, the duplication of official agencies dealing with rural land use and the conflicting advice they often seem to provide, allied to the differing demands of the amenity societies themselves, have all too often led to confusion.

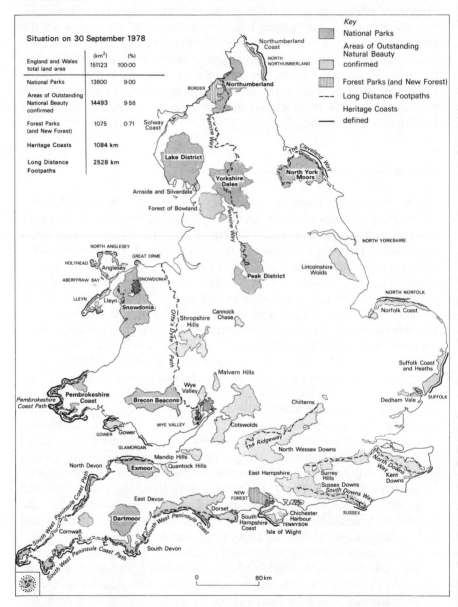

Figure 2.1 Conservation of the countryside. Nearly 50% of the countryside is covered by some sort of landscape protection policy based on a combination of positive and negative powers. (Countryside Commission 1979. *11th Annual Report, 1978*, H.C. 111 (79–80). London: HMSO.)

In view of the difficulties involved in reconciling the competing land use aspirations of official agencies and pressure groups, it is hardly surprising that the government has tried to isolate certain problems by concentrating on particular areas and providing them with special powers. The philosophy and purpose of National Parks, Areas of Outstanding Natural Beauty (AsONB), Heritage Coasts, Green Belts, Country Parks, Picnic Sites, Access Areas, National and Local Nature Reserves, and Sites of Special Scientific Interest, most of which are shown in Figure 2.1, will now be considered in turn.

National Parks

All ten of the English and Welsh National Parks were designated in the 1950s by the National Parks Commission on the basis of the definition in the National Parks and Access to the Countryside Act 1949, 'those extensive tracts of country . . . that by reason of their natural beauty and the opportunities they afford for open-air recreation . . . in which it is especially desirable that measures shall be taken to preserve and enhance their natural beauty and to promote their enjoyment by the public' (Section 5 reordered).

The definition closely followed the advice of the pioneering Dower and Hobhouse reports of 1945 and 1947 (Cmd 6628 1945, Cmd 7121 1947) and gave the National Parks two equal purposes, recreation and conservation. The dual aims caused problems from the outset and the difficulties were further compounded by a requirement that National Park Committees also do their utmost to maintain agriculture and the wellbeing of the local inhabitants and their economy. As pressures on the Parks have grown it has proved increasingly difficult to reconcile these major purposes and the 1970s have seen a reassessment of both priorities and administration.

Initially, all the Parks were administered by Committees of the relevant county councils, except for the Lake District and the Peak District which had their own Boards. It meant that local considerations invariably took precedence over national needs and that little money was specifically allocated to the Parks since the majority of expenditure had to by borne by local rates. In 1974, however, major changes were initiated following the provisions of the Local Government Act 1972. First, the National Parks were given permanent staffs headed by National Park Officers. Secondly, they were given greater planning powers, especially in development control matters. Thirdly, the budget was doubled and the proportion contributed by central government raised dramatically to 75%. Fourthly and, in the long-term, the most significant, each National Park was given the task of preparing a National Park Plan, which was to be a five-year management plan allocating money and resources to specific projects. In addition the Peak and Lake District Parks were required to prepare Structure Plans.

Although the 1974 reorganisation improved administration, it still left open the question of land use priorities in the Parks. Fortunately, 1974 also saw the publication of a major review of National Park policies, the Sandford Report, which recommended that in the last resort conservation should have priority over recreation (DOE 1974d). The government accepted this view in 1976 (DOE 1976d). Deep seated conflicts still remain however and, although a 1976 House of Commons Expenditure Committee Report praised the cost-effective work of the National Parks, it also recommended further reviews of the priorities of National Parks, their boundaries and administration (House of Commons Papers 433 and 256, 1976 and 1977). It is clear from the evidence of both these reports that many issues of principle, let alone detail, need to be settled if the Parks are to resolve the conflict between landscape conservation and rapidly growing demands for further land use change (Smith 1978).

Areas of Outstanding Natural beauty and Heritage Coasts

AsONB are designated by the Countryside Commission under the powers of the National Parks Act 1949 and the Countryside Act 1968 and, by 1976, there were 33 of them, covering 9% of England and Wales (Countryside Commission 1978b). The AsONB are administered by local authorities and policies for them are contained in Structure and Local Plans. These policies normally contain a claim that development control standards will be more rigorous in the AsONB than elsewhere, particularly as regards design control. Theoretically AsONB are of equal landscape status to National Parks, but although grant aid is available for landscape improvement, the funds are even smaller than for National Parks and little effective work had been done in them by the late 1970s.

Heritage Coasts are the result of years of negotiation between the Countryside Commission and local authorities, which began in the late 1960s. They are stretches of undeveloped coastline of high scenic quality and though they may be formally approved in Structure Plans and by the Countryside Commission, they have no statutory basis. The main policy instruments for management are strict development control, and, sometimes, the employment of a Heritage Coast officer to co-ordinate private developments with public investment in the amenities of the area.

Green Belts

Green Belts also have no statutory basis, except that they may be included in an approved Development or Structure Plan. (Housing and Local Government 1962). The first Green Belt was established around London in the 1930s, but the precedent was not followed widely until 1955 when

Ministry of Housing Circular 42/55 advocated their more general adoption. Green Belts in one form or another now cover 15 000 km^2 and are found around nearly all of Britain's major cities. In 1955 their aims were set out as being three-fold:

(a) To check the further sprawl of built-up areas,
(b) To prevent neighbouring towns from merging into one another,
(c) To preserve the special character of a town.

Subsequently recreational use has become an important fourth aim. Development control has again been the main weapon used by local authorities to achieve the first three objectives, while recreational use is incorporated in management schemes, often based on the provision of Country Parks.

Country Parks and Picnic Sites

In a positive attempt to steer recreational pressure away from protected landscapes and to create new recreational foci around towns and thus reduce farm trespass, the Countryside Act 1968 gave grant aid powers to the Countryside Commission to help in the setting up of Country Parks and Picnic Sites. Their main purpose is to provide or improve opportunities for the public to enjoy the countryside, and either private individuals or local authorities may sponsor them, although it is generally easier for local authorities to obtain grant aid. Nonetheless, many private individuals who had traditionally provided access to stately homes or landscaped gardens seized the opportunity to use grant aid in the years immediately following the 1968 Act, although in the 1970s it has been mainly the local authorities that have taken the initiative and created new Parks or Picnic Sites. By 1977 over 140 Country Parks and 180 Picnic Sites had been approved by the Commission for grant aid, providing a valuable additional recreational resource and relieving more vulnerable areas from pressure.

Access areas

In spite of the extra facilities provided by Country Parks and Picnic Sites, demands for recreational use of the countryside have outstripped the new supply of facilities and if protected landscapes are to be conserved, new access arrangements to the wider countryside have to be concluded. Unfortunately, the 1949 Act only gave local authorities and National Park authorities powers to makes access agreements with landowners or tenants on 'open country' and not farmland. This restriction and a general unwillingness to use these powers has meant that only 37 000 ha of land were thought to be covered by access agreements in 1977, even though the powers had been extended to include all rural land in the Countryside Act 1968 (Gibbs & Whitby 1975). In addition to access agreements, county councils must regularly survey and produce a map of all rights of way in their county, and these are now shown on the Ordnance Survey 1:25 000 map series. However, many of these footpaths and bridleways are not very suitable for informal recreation and the Countryside Commission is encouraging rationalisation of the network to provide a more useful resource. One successful initiative has been the creation of the Long Distance Footpaths. Although they took a long time to establish, they have now become so popular that in some places, such as the Pennine Way, they are beginning to suffer from erosion through over-use.

National and Local Nature Reserves

Since 1949 the Nature Conservancy Council has established 153 Nature Reserves of national and international importance covering some 120 000 ha. Out of this total the Council owns 33 000 ha, leases 16 000 ha and manages 72 000 ha on behalf of the owners (Nature Conservancy Council 1977b). When creating reserves, the Council's overriding aim has been to provide a nationwide set of representative habitats in the national interest, primarily for scientific purposes, but with educational and recreational access allowed where appropriate. In addition, the local authorities have set up over 40 Local Nature Reserves and voluntary bodies have created over 1000 reserves of their own. Reserves are, however, by their very nature isolationist, and without the acceptance of the need for conservation in the countryside at large, preservation of habitat within their narrow confines can only be a temporary expedient. Accordingly, the Council are endeavouring to encourage farmers to become more conservation minded and to plant up unproductive land for wildlife habitats.

As yet, existing reserves are far from comprehensive in their coverage. A mid-1970s review by the Council found 700 key sites, covering 950 000 ha or 4·1% of Britain, but only 12% of the land involved was owned or leased as National Nature Reserve (Nature Conservancy Council 1977a). The Council recommended that as many as possible of these sites, particularly the 670 000 ha of Grade 1 status, should become reserves immediately, even if it were necessary to continue farming them until national funds could be found for their purchase.

Sites of Special Scientific Interest

Under the 1949 Act the Nature Conservancy Council can designate as Sites of Special Scientific Interest (SSSIs) areas which need protection because of their flora, fauna, geological or physiographical features, and by 1977 there were over 4000. SSSI designation does not prevent development, it only means that development proposals made to the local authority by developers or public agencies like the Regional Water Authorities have to be submitted to the Council for comment. In line with much planning control the main effect is usually to ameliorate the impact of the development rather than to prevent it, a particular case in point being the growing trend to drain wetland sites like the Somerset Levels and Ribble Marshes, where the Council has no power totally to prevent drainage change. Indeed, in the case of the Ribble estuary the Council could not have prevented any drainage without the aid of a special grant of £1·75 million from the government, given to help the Council purchase the site.

Conclusion

It is all too easy to present an unbalanced picture of the forces at work in shaping the countryside over the past 30 years. In terms of both government aid and private investment, the development of agriculture and forestry, housing and employment have been paramount; the needs of conservation and recreation subsidiary, although their claims have often been strenuously, even stridently prosecuted. Currently direct annual expenditure by central government on conservation and recreation amounts to little more than £10 million, while agricultural support alone costs over £400 million. Since agricultural finance is so tightly controlled by government through the plethora of positive powers, negative controls and advisory services described earlier, and since the industry is such a dominant force in the landscape, it is obvious that government itself has been largely responsible for the nature of the present rural scene through the operation of its policies. Strong arguments are now being advanced that the influence of these policies has been inordinate. While the countryside must be economically healthy if it is to cope with all the demands placed on it, there is evidence that the pendulum has swung too far in favour of resource development, even taking into account the substantial controls outlined above.

3 The nature and measurement of land use change

For many the countryside is one of the last frontiers of individuality and freedom, where the lucky few can still make eccentric decisions far removed from the restrictions and bureaucratic frustrations of urban life. Chapter 2 showed clearly that the freedom is for the most part illusory, but land ownership or control over the rights to use land for farming, fishing, shooting and other activities do confer privileges on many sections of the rural population, denied to most of their urban fellows.

The bulk of rural land is still owned by farmers and they certainly hold most of the rights to its use, and it follows that their motives and attitudes are crucial for a proper understanding of countryside change. Although it would be unwise to generalise too much about such idiosyncratic individuals as farmers tend to be, they may be tentatively divided into five broad groups:

(a) The large estate farmer with business interests or capital outside farming,
(b) The medium size family farmer,
(c) The institutional and public land owner,
(d) The small and often under-capitalised farmer,
(e) The part-time or hobby farmer.

The large estate farmer, the medium size family farmer and the hobby farmer may sometimes still indulge themselves in the luxury of a way of life where change is espoused only slowly, but for the most part change and new technologies are vital to the economic survival of agriculture as an industry and of farmers as a group. Both the structure of farming and the numbers of farms and farmers have altered radically in recent years, with drastic consequences for the landscape. New patterns of land use mean that the landscape of even a generation ago has all but disappeared and the change is likely to continue in the foreseeable future.

The new structure of farming

Until about 20 years ago the overall number of farmers had been relatively stable, largely because of the lack of suitable alternative employment (Gasson 1969), but more recently they have been retiring from the industry and have not been fully replaced by younger entrants. The net result is a shrinking and ageing population, which tends to reinforce the tendency of the farming community as a whole to be conservative, cautious and resistant to unneccessary change. By June 1976 the number of farmers in the UK was officially estimated to be 292 000, but a more realistic estimate (counting only full-time farmers making a significant contribution to output) would be 120 000.

The decline has been greatest in small holdings, indeed there has actually been an increase in the number of larger holdings. Between 1968 and 1975 the number of significant holdings fell from 316 000 to 273 000 in the United Kingdom, leaving 40 000 large holdings (needing four or more full-time workers) compared to 35 000 in 1968, 53 000 medium size holdings (needing two to four workers) compared to 58 000 in 1968, 65 000 small holdings (needing one to two workers) compared to 87 000 in 1968, and 115 000 as very small holdings (needing less than one worker) compared to 136 000 in 1968. Large holdings contribute 56% of production while the very small holdings only 6% (MAFF 1977b). Overall it has been estimated that the number of full-time farmers fell at the rate of 2% per annum between 1963 and 1975, producing a situation where fewer farmers now have even greater power over larger areas of landscape.

The number of farms and farmers and the size of farms have not been the only structural change in agriculture in recent years. Another and longer-term trend has been a transition from rented tenure to owner-occupation as shown in Table 3.1 leading to the position illustrated in Figure 3.1. The change from rented tenure to owner-occupation has had two important effects. First, the constraints imposed by strict tenurial agreements have been removed, and secondly, while tenants have little incentive to improve someone else's property if they become owner-occupiers they have every incentive to invest in improvements. Accordingly, there is now a greater incentive and freedom to change and alter farm landscapes than before.

Table 3.1 Land tenure by holdings and area 1891−1974. (MAFF 1968, 1976. *A century of agricultural statistics* and *Agricultural statistics*. London: HMSO.)

Year England and Wales	Holdings '000		Millions of hectares	
	owned	rented	owned	rented
1891	69	404	1·62	9·72
1921	70	350	2·02	8·50
1960	158	124	5·67	5·67
1974	129	80	6·07	5·26

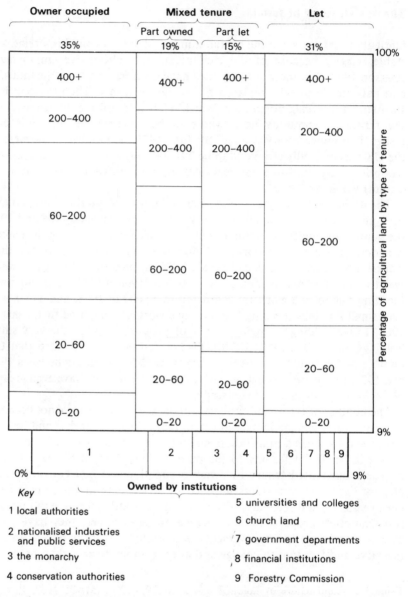

Figure 3.1 Agricultural land tenure in England and Wales 1973 (ha). The figure shows that the medium sized farm, 60–200 ha, remains the most common farm unit. (Country Landowners Association 1976. *The future of landownership*. London: The Association.)

The increase in owner-occupation may, however, be coming to an end as the high price of land, between £1300 and £2400 per hectare in the summer of 1977 (rising to £4000 per hectare in 1980), makes it almost impossible for private individuals to buy and pay for it out of farming income alone

(Cmnd 7058 1978). Their place has partially been taken by financial institution buyers, particularly in the large estates where Capital Transfer Tax has forced the sale of land. Nevertheless, as Table 3.2 shows, their total holdings are still quite small and their significance can easily be overestimated. A study by the Centre for Agricultural Strategy in 1976 showed that government departments accounted for 380 000 ha, local authorities for 357 000 ha, the Crown for 170 000 ha and financial

Table 3.2 Agricultural land ownership in Great Britain 1978. (Cmnd 7599 1979. *Report of the Committee of Inquiry into the acquisition and occupancy of agricultural land*. London: HMSO.)

Category of owner	Area owned million (ha)	Proportion of total (%)
public and semi-public bodies and traditional institutions	1·5	8·5
financial institutions	0·2	1·2
private individuals, companies and trusts (by subtraction)	16·0	90·3

institutions for only 150 000 ha although their share was the fastest growing (Gibbs & Harrison 1974, Harrison, Tranter & Gibbs 1977). It has also been shown that the majority of their holdings are in fact let to tenant farmers so that many of the losses are more apparent than real (Denman 1977).

Such reassurances did not quell the fears that control of agricultural land was being irrevocably taken out of the control of the farming community and, in 1977, the government set up a committee chaired by Lord Northfield to examine the whole question of agricultural land ownership. In their report they urged that land should be regarded as a resource of major national importance and that government policy for rural areas should be directed to four main objectives:

(a) An efficient agriculture,
(b) The retention of personal commitment and incentive,
(c) Conservation and preservation of the countryside,
(d) Viable rural communities (Cmnd 7599 1979).

After sifting through 350 written submissions and a mass of oral and other evidence, the Committee produced a picture very similar to that already published by the Centre for Agricultural Strategy as shown in Table 3.3 and concluded that changes in land ownership were unlikely to threaten their four main policy objectives. They predicted that land would remain predominantly in private hands and that the proportion owned by financial institutions by 2020 would not be more than 7% of the present agricultural area of Great Britain. Even their most pessimistic assumptions only produced a holding of 11% by the financial institutions. Nevertheless, the

Table 3.3 Agricultural land ownership by public and semi-public bodies and traditional institutions (ha). (Cmnd 7599 1979. *Report of the Committee of Inquiry into the acquisition and occupancy of agricultural land*. London: HMSO.)

	Cmnd 7599 (1977)	*CAS data (1976)*
central government*	462 421	364 220
local authorities	364 967	356 639
the Crown	163 567	168 900
statutory agencies and nationalised industries	225 040	221 445
conservation bodies†	132 164	103 609
educational establishments	97 516	86 588
religious institutions	69 626	68 863
total	1 515 301	1 370 264

Note: The discrepancy is largely due to the fact that Cmnd 7599 received fuller information from the Ministry of Defence.
*Includes Forestry Commission land in agricultural use.
†Private organisations, National Trust and RSPB.

Committee did suggest certain modifications to taxation and tenure legislation so as to slow down the decline in the rented sector, which it saw as a vital component in the fabric of farming. In this they echoed the recommendations of others and it certainly seems to be true that small tenanted properties, suitable for those entering the profession without the benefit of land already in the family, are becoming increasingly scarce (Walston 1978).

Whoever owns or rents a farm holding, the income expectations are fundamental to understanding why farming methods are changing. Each of the possible major farm enterprises involves different degrees of capital investment and labour input, and have different risks and profit margins. At present, milk production is widely favoured since returns are monthly and are as certain as anything can be in an industry so susceptible to outside forces like the climate. On the other hand, pig production is a high risk business where profits or losses can be spectacular and very unpredictable. Most farmers having settled for one of the basic enterprise groups (see Table 3.4) with their differing patterns of profit, machinery, skills and commitment, are generally loth to switch to another group unless there is a really pressing financial need when times become hard. They normally prefer to intensify existing production by actions such as liming and fertilising fields, draining wet land, reclaiming scrub and heathland, and removing hedgerows and trees, thus producing the land use and landscape changes that have so worried planners, conservationists and the general public.

The need to intensify and specialise production of the two major farm enterprises, namely the production of grain for feedstuff and the production of meat and milk products from cattle, now dominates the

Table 3.4 Specific net incomes for different farm enterprise groups. (Cmnd 7058 1978. *Annual review of agriculture 1978*. London: HMSO.)

Type of farm	Average size in ha	Standard man days	Average net income £	
			1975–6	1976–7
specialist dairy	50	970	6 247	6 508
cattle and sheep	83	759	5 982	6 691
cereals	142	990	9 489	12 474
pigs and poultry	47	1 200	12 788	9 625
general cropping	92	1 254	16 746	17 875

Note: 1975–6 is distorted due to the drought of 1976.

future of the agricultural industry. The two systems are in fact closely interrelated, for as Table 3.5 shows, the main income of British farmers is derived from animal products while the main item of expenditure is animal feedstuffs. This integration of production systems conveniently divides the British Isles into an arable east and a pastoral west and, because arable farming is more responsive to intensified inputs, explains why most of the major land use changes occurred first in the east and then more slowly and on a smaller scale in the west.

Table 3.5 Balance between farming revenue and expenditure 1977 (forecast). (Cmnd 7058 1978. *Annual review of agriculture 1978*. London: HMSO.)

Revenue	£ million	Expenditure	£ million
milk and milk products	1 461	feedstuffs	1 770
pigs and poultry	1 102	machinery	980
fat cattle and calves	1 046	labour	935
grain	788	miscellaneous	469
horticulture	704	fertilisers	436
other crops	585	maintenance	245
eggs	393	seeds	204
sheep and lambs	283	livestock	138

Size of farm is another important factor in explaining the motivation behind changes in land use. Average efficiency increases markedly with farm size up to about 60 ha, more slowly up to about 160 ha, but its progress thereafter is erratic (Britton & Hill 1975). As Table 3.6 shows, arable farms are mostly in the size range where efficiency increases are most closely related to intensified production, an added reason why landscape change has been the greatest in the arable east. Since the average size of holding in 1975 was only about 45 ha, there is still a great deal of scope for further increases in efficiency and, consequently, land use change (MAFF 1977b).

The trend towards larger farm units would no doubt have been greater if small farmers did not remain in the industry long after their farms had become uneconomic. When such a farm is finally given up, it is very often amalgamated into an adjacent or nearby holding which has the capital, skill, labour and entrepreneurial enthusiasm to make the changes that the previous occupier lacked. At the same time the holding will usually be rationalised, by increasing field sizes, replacing worn out buildings with one central structure, and by generally tidying the place up, producing a landscape aesthetically pleasing to the farmer in search of efficient ordered beauty, but boring and monotonous to the town dweller and conservationist in search of their 'rustic idyll'.

Farm finance is also an important element in countryside change. Agriculture with £28 000 (close to £100 000 by 1980) invested in machinery and fixed equipment per head of the labour force in 1976 has a greater capital input per worker than even such capital intensive industries as the chemical industry (National Economic and Development Office 1977). The huge capital investment is largely financed out of income, rather than borrowing and helps to explain the pressure to maximise production and to make the most use of farmland and machinery by adapting the landscape to fit the expensive machines. It can be argued that the very prosperity of the industry has been the main initiator of the massive land use changes; it is no accident that the 1930s, a period of extreme depression in farm incomes, witnessed few of the land use changes seen in post-war times even though new machinery was also being introduced in that period. The financial success of the post-war years has however been constantly threatened by the increased impact of both current taxation and capital taxation at death (Evans 1969, Waller 1976), with the spectre of reduced incomes and shortages of funds for investment. Some farmers have reacted to this by

Table 3.6 Numbers and size of holdings by enterprise group. (Cmnd 7058 1978. *Annual review of agriculture 1978*. London: HMSO.)

	1977 (provisional)
Crops and grass	
0·1 to 19·9 ha	112·0
20·0 to 49·9 ha	71·4
50·0 to 99·9 ha	41·5
100·0 ha and over	29·4
Cereals	
0·1 to 19·9 ha	71·3
20·0 to 49·9 ha	22·5
50·0 ha and over	21·3
Dairy cows	
1 to 29	33·1
30 to 59	22·2
60 and over	18·7
Beef cows	
1 to 19	66·4
20 to 49	18·5
50 and over	8·2

investing unnecessarily heavily in powerful and expensive new technology in order to offset these costs against taxable profits. Whatever the justification, their actions have certainly accelerated the need to change the structure and appearance of their farms to accommodate the new machines.

The availability of labour has also contributed to the nature and pace of change in the countryside, for the intimate landscape of the past cannot be maintained without a large workforce. A shortage of labour leaves no spare time for landscape conservation and leads to the substitution of machinery wherever possible and thus to an easily managed and stereotyped farm landscape with as few unnecessary features as possible. Although in the world at large 50% of the labour force is still engaged in agriculture, the figure in Britain has now fallen to only 2% (Grigg 1976). Even if all those engaged in ancillary industries are included the figure only rises to 10% (Countryside Review Committee 1978). There are probably only 200 000 full-time workers left, and 75% of these are employed on large farms with more than four employees. Many smaller farms have no workers and rely on neighbours or contract labour to help out at peak times. Indeed, the growth of contract labour and machinery further undermines the possibility of full-time workers managing the landscape for non-agricultural purposes, and is accompanied by a loss of craft skills like hedge laying as the traditional labour force declines. Without doubt, the lack of labour has further encouraged the switch to simpler more intensive farming systems away from mixed farming, creating a less diverse landscape.

The exodus of farm workers has also left a significant gap in the social life and economic viability of services in the countryside. Thousands of tied cottages have already been released for occupation by other people (Gasson 1975, Irving & Hilgendorf 1975). The provisions of the Rent (Agriculture) Act 1976 which gives security of tenure to workers could further accelerate this trend as farmers become less willing to tie up a cottage in this way (Rossi 1977). The Act will almost certainly increase the desirability of using contract labour and exacerbate the shortage of resident staff needed to maintain the wider fabric of the countryside.

In summary, the farming industry is still dominated by farmers born and bred to the industry, even if there are fewer of them, and to understand why they might wish to change the countryside it is necessary to appreciate their

Table 3.7 Cambridgeshire farmers' criteria of 'good farmer' by farm size. (Gasson, R. 1973. Goals and values of farmers. *J. Agric. Econ.* **24**, 521–37.)

Good farmer is one who:	Size of farm		
	small	medium	large
produces best crops or livestock	42	36	46
leaves the land better than he found it	40	41	33
is progressive, up-to-date, experimental	23	37	32
preserves the beauty of the countryside	21	11	11
is making most money	10	13	19

goals and values. Surveys have shown that what farmers value most of all is the independent open-air way of life (Gasson 1973), for farming is much more than merely a job and the idea that farmers wilfully change the countryside without good cause and much thought is patently untrue, as revealed in Table 3.7. Changes have been as much forced onto farmers by outside pressures as voluntarily espoused, for farming is not a self-contained industry and although internal alterations to its structure go a long way to explaining land use changes, external factors like the demands of consumers and the food processing industry, the advent of new technology, and the government interference referred to in Chapter 2 are very important external factors.

The consumer and the food processing industry

About 25% of all consumer expenditure is on food and about a third of this finds its way back to the farmer, giving the consumer a powerful direct influence on the farming industry. In Britain, people have come to expect a cheap and plentiful supply of food and the politically powerful voices of the housewife and the food processing industry have done much to ensure that the public has enjoyed low-cost and regular food supplies for most of this century, apart from the periods of wartime shortages. Traditionally, the British farmer has had to contend with severe overseas competition, for until entry into the European Community in 1973, Britain imported nearly half her food from low-cost suppliers in the Commonwealth and the Third World. Since then the eight other members of the Common Market, many of them with surplus produce, have stepped in and assumed the role of food exporter to Britain as shown in Table 3.8.

The long-term result of the flow of competitively priced food imports into Britain and of consumers' demands for low prices, has been an emphasis on low-cost production. There has been continuous pressure on farmers to intensify production systems by removing redundant landscape features, draining wetlands and reclaiming moorland, and generally trying to extract the largest possible yields from every available hectare.

A continuation of these trends depends on the further evolution of the British diet and the degree to which surpluses of food outside the UK continue to be available. It has been argued that if the British diet were to return to a simpler one based on potatoes, cereals and milk products, and eliminated the wasteful production of food based on meat and animal products which generally are poor converters of arable protein into animal protein (Mellanby 1975), Britain could be self-sufficient. Another factor, the availability of world food surpluses, could also alter current expectations for as a 1976 government statement forecast, 'Because competition for surplus supplies of food on the world markets will increase as population and requirements grow and because of increasing costs of

Table 3.8 Imports and self-sufficiency in cereals and meat. (Cmnd 7058 1978. *Annual review of agriculture 1978*. London: HMSO.)

Total cereals '000 tonnes	Average of 1967–9	1977 (forecast)
UK production	13 773	16 963
imports from the Eight	950	3 044
imports from third countries	7 468	5 620
production as % of new supply	64%	68%
Total meat supplies '000 tonnes	Average of 1967–9	1977 (forecast)
UK production	2 506	2 787
imports from the Eight	469	499
imports from third countries	617	293
production as % of new supply	71%	83%

production in real terms the cost of food imports is likely to rise' (Cabinet Office 1976, p. 20). Whatever happens it is likely that pressure on farmers to produce more food will continue and the trend towards simplification of the present multiple land use patterns will be sustained.

The rapid growth of food processing has also deeply influenced farming methods and the landscape. As Table 3.9 shows, manufactured food has shown the fastest rate of growth in recent years and in just one year sales of frozen convenience food rose in real terms by 11%. Furthermore, food processors' purchases of all agricultural output rose from 22% in 1972 to 48% in 1976 giving the processors a powerful voice not only in what is grown, but also in how it is grown (Ashby 1978). Many crops are now contracted to a food processor even before they are sown, and their subsequent growth is closely controlled by him rather than by the farmer. The type of quality control exerted by major firms like Marks and Spencer and Birds Eye allow for few if any blemishes on crops and demands the ruthless elimination of all pests, diseases and predatory wildlife.

Ironically, the product is then often marketed with romantic packaging and advertising depicting old-fashioned production methods giving a very different image of the countryside from the harsh reality. The misconceptions that this deception can cause are well illustrated by a 1973 study of the attitudes of urban dwellers to farming (National Farmers Union 1973). It showed that while 73% believed that farmers were adequately preserving the countryside, 61% thought that too many

Table 3.9 Consumers' expenditure on certain foods. (Central Statistical Office 1977. *Social trends 1977*. London: HMSO.)

At constant prices (1970 = 100)	1961	1971	1976
bread and cereals	102	97	95
dairy products	90	103	107
meat and bacon	92	100	97
manufactured food	86	94	111

chemicals were used and 58% disliked the intensity of modern food production. The confusion of attitudes neatly picks out the central dilemma facing rural planners; the need to balance economic progress with preservation of wildlife and attractive landscapes. However, there is little doubt that economic forces will continue to hold sway and that the growing demands and control of the food processing industry will lead to yet further intensification of agricultural systems and an even greater loss of wildlife, landscape quality and ecological diversity.

New agricultural technologies

The rapid post-war development of new technologies, particularly in machinery, buildings and chemicals has had two main effects on agriculture. First, the replacement of labour and horses as shown in Figure 3.2, and secondly, the ability to transform the scale of operations from the human scale to that of the machine. The revolution has so far been one of degree rather than of kind, for traditional processes have been mechanised rather than fundamentally rethought (Smith 1975). If and when they are, the changes of the post-war period may seem minimal in retrospect.

To obtain a satisfactory return on the considerable investment in machinery and chemicals, non-productive time must be minimised and it has been calculated that square fields of about 20 ha are the most efficient for modern machinery to operate (Edwards 1969). These are now rapidly replacing the irregular, smaller units inherited from previous systems, though in eastern England, larger geometric fields created during the eighteenth-century enclosures were already quite common. More powerful machines have also allowed the exploitation of land formerly too wet or steep for cultivation. In addition the savage, indiscriminate ferocity of new hedging and ditching techniques has prevented the growth of hedgerow trees and the survival of habitats in these important, traditional linear features of the rural scene. A side effect of the size of modern machinery is the replacement of small farm buildings, with single purpose, uniform structures constructed from universally available materials. This has had a marked effect in many areas on the look of the countryside and, as will be shown in Chapter 6, is almost totally outside present planning controls.

Energy, pesticides and fertilisers are now mainly derived from oil and are fundamental to the success of modern farming systems. By reducing the incidence of pests and diseases they have made a major contribution to the huge increases in productivity achieved in the post-war period, but at the same time they have led to the loss of so many species that the more dangerous chemicals are now controlled because of their ecological effect. Nevertheless, there are some reasons for supposing that much of the increase in productivity may be more apparent that real. The replacement of the horse by modern machinery has freed the 1·3 million ha formerly

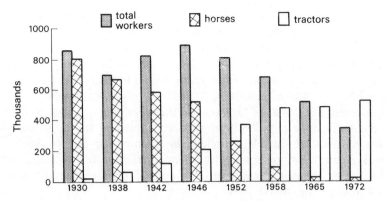

Figure 3.2 Workers, horses and tractors in Great Britain 1930–72. Workers and horses have been replaced by tractors as agriculture has become increasingly mechanised. (MAFF 1968, 1973. *A century of agricultural statistics 1866–1966* and *Agricultural statistics* 1972. London: HMSO.)

devoted to the production of oats for horse fodder, and, in any case, the output has been achieved through the heavy use of energy, pesticides and fertilisers and may really represent a net loss rather than a net gain in terms of calories produced (Juniper 1977, Mellanby 1978). In spite of these long-term doubts, it is certain that in the short-term, farmers have much to gain from applying modern technology to their farms, even though modern farming systems rely heavily on inputs of energy derived from oil.

The net result of these new techniques has been a swing away from the rotational practices, formerly necessary to preserve fertility and reduce pests and diseases, towards an increasingly dominant monoculture. Another change has been an increase in the area of drained land from 50 000 ha in 1950 to 250 000 ha in 1972 (Carter 1973). Draining and new ploughing techniques have also allowed a great increase in the area of

Table 3.10 Agricultural production and productivity. (Cmnd 7058 1978. *Annual review of agriculture*. London: HMSO.)

Product	Unit	1967–9 (average)	1977 (forecast)	Average yield 1967–9	Yield 1977 (forecast)
wheat	'000 t	3 579	5 280	3·91	4·92*
barley	'000 t	8 717	10 745	3·61	4·45*
potatoes	'000 t	6 763	6 571	24·90	28·60*
eggs	million dozen	1 231	1 163	210·50	240·50†
milk	million l	11 908	14 415	3 673·00	4 407·00‡
beef	'000 t	1 202	1 206	—	—
poultry meat	'000 t	515	646	—	—
pork	'000 t	583	631	—	—

*t/ha	†eggs/bird	‡l/cow

temporary leys at the expense of the floristically richer permanent pastures, and finally zero grazing, where the grass is cut and taken to the livestock, is making hedges and fields redundant in pastoral as well as arable areas.

The post-war period has seen the evolution of a very productive agriculture with three or even four-fold increases in yields as shown in Table 3.10. The cost of these improvements has been paid elsewhere: in the loss of livelihood for thousands of farmers and their workers; in a loss of diversity and self-sufficiency as farms become more specialised and thus dependent on the vagaries of outside suppliers; in a simplified and probably less attractive landscape; and finally, in a loss of freedom for all concerned as the increasing complexity and severity of outside forces reduces the farmer's ability to chart his own destiny.

The changing farming scene: a summary

The main conclusion to be drawn from the preceding analysis is that farmers remain the single most important agent for determining the degree of land use change in the countryside. However, their influence has declined as they have been increasingly subjected to the outside forces of economic and technological change, leaving them with less scope to influence the degree of any change. Moreover, government, both national and European, is further reducing the farmers' freedom of action. Nonetheless, they still have considerable power to alter the details of land use patterns according to their own whims and to radically alter the nature of the countryside within a single calendar year if they so wish. This would, however, be totally out of character as the majority of farmers react slowly to change and have not espoused the new husbandry of post-war times rashly or without deep thought. The scientist, the food processor, the cheap food lobby and the government are just as much the architects of post-war farming change as the farmers themselves and although a countryside run by agricultural scientists and agricultural economists might just possibly be more productive than the one we have, it would also be far less diverse and interesting, lacking the conservation ethic that many farmers practise, either out of concern for their stewardship of land or to preserve their sporting interests.

Despite these outside pressures, it is the individual farmer who still has to make the decision to alter his land use. Just how and why does he come to a particular decision? The main factors affecting attitudes to change are age and education, whether a farmer owns or rents the farm, the desired level of income and the risk each individual is willing to take to achieve it, size of farm, the availability and cost of finance, the level and stability of government aid and price support, the rate of taxation, and, finally, the likelihood of the farm remaining in the family when a farmer retires.

In combination, these factors have led to the major changes in the structure of agriculture during the post-war period identified earlier: increasing mechanisation, reductions in farm manpower, amalgamation of holdings, concentration of production by area, specialisation by individual farms, and modifications of traditional husbandry techniques (Davidson & Lloyd 1978). These trends are likely to continue as world food surpluses diminish, although surpluses within the EEC may temporarily undermine the need to further increase production and may shift the balance of advantage away from arable towards pastoral farming for a time.

Finally, it is important to note that agricultural change has not occurred evenly. In addition to the regional differences already noted between the arable east and the pastoral west there are significant local variations where physical conditions of land quality, slope, and climate impose a different choice of enterprise from the regional norm (Coppock 1974). Furthermore, the location of farmland in relation to centres of population also exerts a strong influence on the way it is used, and remote areas are unlikely ever to experience the same kinds of pressure that exist, for example, in the urban fringes. In other words the physical condition of land and its location provide a deep underlying stability in the agricultural use of the countryside. The changes of the last 30 years, although they may seem dramatic, are but another layer in a gradual sequence of change that has caused the landscape to evolve through a number of different phases in the past 1000 years.

New developments in forestry

Forestry is not a homogenous industry. It is divided between the private and public sectors, between lowlands and uplands and between hardwoods and softwoods. During this century it has spread rapidly and the proportion of land under timber has risen from 5 to 9%. The planting rate accelerated from 30 000 ha per decade in the 1920s to 360 000 ha in the 1960s (Forestry Commission 1974a). Since the war, private owners have also been planting more trees and have added significantly to the efforts of the Forestry Commission (Ch. 2).

Most private planting was historically mainly for local purposes and confined to land too steep, wet or exposed to farm. As a result the majority of new plantations were small and scattered and could never have satisfied the needs of a rapidly expanding industrial nation. However, many of these small woods still today provide a visual and ecological resource of enormous value, especially in England where they account for 72% of the total. They are less significant in Scotland where twentieth-century coniferous plantations now dominate the scene.

Unfortunately, many small woodlands are now under threat because modern economic pressures have made small-scale forestry uneconomic and were it not for grant aid many of the trees would have already been felled.

Even those plantations that remain are usually badly managed and unproductive (Countryside Commission 1979b), for few contemporary farmers can afford to tie up large sums of capital in trees that will not mature for at least 60 years in the case of conifers, or well over 100 years in the case of broadleaves, such as oak and elm.

Post-war afforestation is considerably different from that previously practised. The main demand is now for softwoods, because they are the fastest-growing species and the easiest to use, and planting is now only feasible if economies of scale can be achieved in large plantations where the investment in skilled labour and expensive machinery can be justified. As a result, coniferous plantations now dominate in strict contrast to the situation in the decade 1890 to 1900 when private owners planted 30 000 ha of broadleaved woodland and only 5000 ha of coniferous woodland. Indeed in the decade 1960 to 1970, the situation was totally reversed and private owners planted only 10 000 ha of broadleaved trees compared to 140 000 ha of coniferous woodland (Forestry Commission 1974a), reflecting the timber processing industry's demand for softwoods (Forestry Commission 1974b).

In line with a switch to large coniferous plantations, forestry has experienced a similar technological transformation to agriculture, with selective breeding of new tree species, the applications of pesticides, herbicides and fertilisers (often from the air), and the introduction of one species monoculture in contrast to the mixture of the unplanned woodland of the past (Dahlstein 1976). Furthermore, specialised management skills and labour are available from forestry consultancy firms like Fountain Forestry or the Economic Forestry Group. In spite of all these advances private forestry is still only marginally profitable, even though Britain still has to import 90% of its timber.

The profitability of forestry as opposed to agriculture depends on the location of the land. In the lowlands the balance of advantage is nearly always towards agriculture, except in areas like the poor sands and gravels of the New Forest and the Brecklands, and on steep slopes or wet sites where agriculture may be difficult. Little, new, private planting has occurred except for amenity purposes although the Forestry Commission has planted up large areas of the Brecklands and other sandy areas. Existing woods are often poorly managed and the low grade timber they contain is often worth less than the cost of clearance, a considerable disincentive to its better management.

In the uplands, however, the situation is rather different. Land prices are lower, because hill farming is much less profitable and predictable than agriculture in the lowlands and there is a real choice between agriculture and forestry, particularly for the landowner with capital reserves and a high tax liability. Traditionally, forestry and agriculture have been seen as competitors for the use of hill and mountain land but, partly because the two are now so finely balanced, they are being reassessed as complementary rather than competitive uses. For example, in some upland estates the two have been integrated with encouraging increases in financial returns, and on a large estate near Fort William the following three advantages have accrued:

(a) Provision of new access roads, permitting better cultivation and easier stock movement,
(b) Division of land into more manageable units,
(c) Shelter of stock and grass leading to better growth rates (Newton 1977).

With the advantages of integration becoming increasingly obvious there is little doubt, that with continued government support several million hectares of upland Britain will be usefully planted, for it has been estimated that Scotland alone has a potential 1·7 million ha capable of producing economic crops of timber (Hamersley 1974, Locke 1976). However, the long-term nature of returns from forestry and government doubts about the wisdom of continuing with a high rate of afforestation may deter some landowners from embarking on what, in the short term, is a somewhat risky investment. The Forestry Commission too is facing difficulty as land prices in the uplands are rising sharply, reducing the amount it can afford to buy, although it has not yet been forced to resort to the compulsory purchase powers that it ultimately has at its disposal.

For the purposes of this book, hedges and hedgerow trees can also be considered part of the forest scene and both have suffered very large losses since the war, most notably in the eastern counties of England where arable farming dominates. Estimates of a loss of 1 million km of hedgerow or a rate of 0·73% of the total per year for the period 1946 to 1970 have been made (Caborn 1971). Hedges are removed mainly because they obstruct machinery, but also because they reduce the amount of cultivable land and harbour pests and diseases. In addition many hedges are allowed to deteriorate and those that remain are trimmed by indiscriminate flail machines, which destroy saplings and reduce hedge diversity, allowing only the hardiest species to survive. The rate of removal has slowed down since 1970 as most of the obviously redundant hedges have been grubbed up and as the positive attributes of hedges for ameliorating climatic effects have been rediscovered. Nonetheless, an important visual and ecological asset has been lost in the eastern counties and remains under threat in the Midlands and the western counties.

Other uses for rural land

Although dominated by agriculture and forestry, the countryside is also used for a wide range of other activities, such as water supply, mining, defence training, transport, recreation and conservation, all of which have produced important localised changes in recent years. Most are the direct result of the burgeoning demands of the growing urban population and are set in rural areas, because it is there that there are extensive tracts of relatively cheap land, uncluttered with settlement.

One of the fastest expanding of these uses is water supply. Demand for water has grown five-fold since 1960 and presently stands at $17\cdot2$ million m³ a day, one-third of it consumed by industry and the rest by domestic households. There is still a surplus of potential supply over demand (Central Statistical Office 1977), but there are problems of exploitation caused by, among other things, the pollution of many lowland supplies by agricultural and industrial effluents, and the fact that most of the best new sources of supply are in the western uplands and remote from centres of population.

For over 100 years the most important water management technique has been reservoir construction (Tanner 1976), and there are now over 540 reservoirs, covering an area of 20 000 ha. Though the total amount of land directly affected is small, the individual sites exercise a disproportionately large influence on the landscape. Their main purposes of supplying potable water and regulating river flow are being added to and now include such subsidiary objectives as recreation, wildlife conservation and field sports. Indeed, there are very few stretches of water in the country that are not in multiple use.

The demand for minerals in Britain has also increased dramatically since the war and this is reflected in greater domestic production. The only important exceptions are coal, iron ore and slate where substitution and cheap foreign imports have caused an overall decline. Even though much lower grade minerals are now capable of economic exploitation, most of the nation's mineral needs still have to be imported, not least because only large and extensive deposits are considered economically viable (Blunden 1975). This has important landscape implications, for it increases the impact of individual mining operations on the surrounding countryside.

There are three particularly important changes in the pattern of mining that need to be noted. First, environmental impact is now very considerable and, since most deposits and most exploitation takes place in the uplands where the bulk of landscape preservation areas have been designated, there are considerable conflicts between the needs of the national economy, local job creation and conservation. Secondly, around most large towns and cities the demand for sand, gravel and limestone, the raw materials of concrete, that most ubiquitous of modern building materials, is so great that large areas have had to be devastated (Jones 1973). Thirdly, the world oil crisis of the 1970s has hastened the development of North Sea oil and, hence, many ancillary coastal developments; and the proposed development of two vast coal mines on previously unspoilt land in the Vale of Belvoir in Nottinghamshire and the Selby area in Yorkshire (North & Spooner 1978, Pearlman 1978).

Although the armed forces released a good deal of land particularly in the immediate post-war period from military use, the Defence Estate still covers an area of about 300 000 ha of which about 250 000 ha lie within National Parks, Areas of Outstanding Natural Beauty or along the coastline (Cmnd 5364 1973). Although the use of this land for defence training is clearly in conflict with its recreational use, it is ironically more compatible with wildlife preservation and to some extent with landscape conservation, because it deters pressure from other potential users. Nonetheless, there have been repeated demands for the release of this land even though it would merely transfer the problem of military training elsewhere. A 1971–3 review of the Defence estate recommended the release of 12 400 ha, improved access at 57 sites, the release of 26 km of coastline and improved arrangements for amenity and conservation (Cmnd 5364 1973). Of the sites suggested, Lulworth Cove in Dorset, and Dartmoor were singled out for special attention but both remain firmly in army hands although concessions for greater outside use have been made. In the case of Dartmoor the difficulty of obtaining an alternative site has proved a major obstacle and even after a public inquiry the government has only agreed to the release of a small part of its holding, so that the basic conflict as to whether southern Britain's relatively scarce wild and open land should be used for defence or recreation remains (Cmnd 5714 1974).

Transport uses of the countryside have grown rapidly in recent years, the most notable incursions being new roads and airports. Not only has the number of vehicles grown from 7·0 million (3·5 million cars) in 1955 to 17·8 million (14·0 million cars) in 1976, the distances travelled have also increased from 80 thousand million vehicle km to 250 thousand million vehicle km over the same period (British Road Federation 1977). To cope with this, expenditure on roads rose from £111 million in 1956 to £1600 million in 1976, resulting in the construction of 2200 km of motorway and a considerable amount of road widening. Since most of this has been in rural areas, the countryside as seen by the majority of visitors, who are car passengers, has been drastically altered in spite of attempts, particularly recently, to ameliorate the impact of new roads on the landscape.

The new roads have also had a wider impact. Motorways in particular and also a number of new trunk roads have opened up previously remote and unspoilt landscapes, although many of these new roads have been bitterly opposed in long drawn out public inquiries. Nearly all of central southern England is now within commuting distance of major cities, and other rural areas are now within the easy reach of weekend or day trippers. Their construction has led to a loss of farmland and the severance of formerly economic farm units into marginally viable fragments (Rural Planning Services 1978). Finally, the axes of the new roads have created corridors of development pressure with foci of more intense pressure at major intersections.

A crucial secondary consequence of improved communications has been the growth of informal countryside recreation, so that it is now the most popular of all the non-home based recreational activities except for social drinking! Surveys in south-east England have found that in the week prior to interview 9% of respondents had made a recreational trip to the countryside, a further 9% had been to the coast and 4% to stately homes (Countryside Commission 1973). In a wider survey no less than 41% of those asked had made a trip to the countryside in the previous month and over 76% had visited the countryside for recreation at some time during the previous year (Office of Population Censuses 1973).

There is known to be a close relationship between high rates of participation in countryside recreation and social and economic position. In crude terms, the more affluent and better educated a person, the more likely he or she is to follow these kinds of leisure pursuits. In recent years, as is well known, social trends have been moving strongly in favour of greater participation in countryside recreation. The rise in car ownership has already been referred to and working hours have fallen from an average of 44 per week in 1950 to 40 in 1974, leaving the kind of time budget available for recreation that is illustrated in Table 3.11.

Other factors, such as stage in the life cycle, are also important, for as Table 3.12 shows, participation reaches a peak for parents with children of school age. Countryside recreation would appear, therefore, not only to be

Table 3.11 The use of time 1974 (h). (Martin, W. and S. Mason 1976. Leisure, 1980 and beyond. *Long Range Planning* **9**, 58–65.)

Essential activities	Per week	Leisure	Per week
sleeping	58·0	in working week	39·3
eating	10·5	in holiday week	86·0
hygiene, shopping, etc.	13·5		
total	82·0		
		travel to work	5·0
Work	41·7	*Total work + essential activities*	128·7

Total work + essential activities per year   6450
Leisure per year   2286

family based, but also orientated towards children. The typical outing is taken in a car where the family act as a unit, in contrast to most other sport and leisure pursuits which are usually undertaken singly.

Finally, geographical location and the opportunity to take part are obviously crucial; inner city dwellers and those living in northern industrial areas are much less likely to be able to enjoy countryside recreation than those in the more suburban south.

The increase in countryside recreation can only be described as explosive. Visits to all types of countryside, but especially National Parks, Country Parks and historic monuments and buildings have at least trebled since the

Table 3.12 Countryside recreation trips in the week prior to interview. (Countryside Commission 1977. *Digest of countryside recreation statistics 1976.* Cheltenham: The Commission.)

Family status	% participating
young single	8·5
young married no children	7·4
married pre-school children	10·5
married school age children	12·3
married above school age children	10·4
elderly married	9·2
elderly single	10·1

war. The sheer weight of visitors has led to conservation problems as sites are either physically eroded away (in a few extreme cases only), or become unpleasantly overcrowded.

The new recreational pressure together with changing agricultural practices and urban growth are the main factors behind the pressure for stronger conservation measures in the countryside. Without effective conservation many recreation sites would soon lose their attractiveness, although it must be stressed that recreation and conservation are not usually intrinsically incompatible and are often closely interdependent (Budowski 1976). Their co-existence is only possible however at robust sites; fragile environments or rare resources must be conserved by the exclusion or restriction of recreational access.

Countryside conservation without recreational use may seem at first sight somewhat pointless but there are other reasons for conservation, notably the ecological desirability of maintaining as diverse a countryside as possible. Plants and animals protected within nature reserves or by law are vital raw material for continued genetic improvements to existing crops and livestock. Furthermore, wildlife in the countryside can often act as a balance to the monoculture of agriculture and provide early clues about the onset of new pests and diseases, in a not dissimilar way to canaries smelling gas in coal mines. In the last analysis, the knowledge that certain rare species still exist, even if they cannot be widely seen, is a comforting thought for urban man only dimly aware of his capacity for destruction.

Social changes in society as a whole have thus had a marked influence on the countryside and on the way it is evaluated as a resource. Affluence and automation have so reduced the number of hours spent in work and household chores that for the first time in history large sections of society have the time and money to look around for pursuits to fill the empty hours (Martin & Mason 1976). The new pressures have led to a realisation that the carrying capacity of the countryside needs to be much more carefully and scientifically evaluated but, just at the time when they were needed, the techniques for doing this were found to be woefully inadequate.

Measuring land use change

After three decades of rural planning, very little is known still about the detailed nature of land use change in the countryside and the effect of planning and other government powers on the changes there have been. Lack of data, the length of time required to collect them, and the difficulty of analysing and presenting the material provide three good reasons why land use change in the countryside has been relatively neglected as a research topic in recent years. Nonetheless, an analysis of the pattern of post-war land use changes probably provides the most cost-effective guide to the impact of rural planning powers on the landscape, if only the three problems referred to above can be overcome.

From national statistics

The most valuable source of nationally collected data are the annual agricultural census figures. They have been used in many studies to examine land use change at the national, regional and local levels, even though the data are not strictly equivalent to land use statistics (Peters 1970). There are also other national surveys of land use but these are either too sectional or irregular to be of much use. For example, the Forestry Commission's occasional census of woodlands only provides data on forestry. Furthermore, the more wide ranging surveys of the Ministry of Agriculture are very infrequent and there has been no major survey of detailed farming land use since the 1943 National Farm Survey of England and Wales (MAFF 1946). The regular regional surveys of sample farms conducted by University Departments of Agricultural Economics, on behalf of the Ministry of Agriculture, unfortunately have an economic bias and moreover are conducted on an inadequate spatial sample to provide data on small-scale land use changes. The annual agricultural census has been conducted every year since 1866 (MAFF 1968). The entire country is covered in early June and samples of certain data are taken in the spring, autumn and winter. The basis of the census is agriculture and so the unit of enumeration is the holding not the farmer. Although the form and content of the questions have changed over the years, three basic types of information are sought. First, land use by crop area, secondly, livestock by totals, and thirdly, the type and number of farmers and workers. The data are published at the county and national scales by HMSO, but detailed data for parishes and ADAS districts can be obtained from the Ministry of Agriculture. In addition, ADAS publishes type of farming maps, which provide a detailed picture of the sub-regional pattern of agriculture.

Despite the apparent detail and widespread coverage of the census, its usefulness for assessing national or local land use change is severely limited. It is collected primarily to help the government formulate national agricultural policy and is accordingly geared to the information needs of the

Annual Review system that requires estimates of changing production totals and stock numbers, not land use change data (Orton 1972, Napolitan 1975). There have been changes in the size of holding included with a trend towards the progressive disappearance of small farms below 2–4 ha, or those requiring less than one man to work them. The amount of agricultural land was also seriously underestimated until the 1930s Land Utilisation Survey and the wartime National Farm Survey revealed how much land farmers had failed to record. The census is not a direct record of land use change, which can only be inferred from changes in the gross totals of each land use, and it therefore under-records two-way changes in the area devoted to certain crops. Another shortcoming is that no motive for change is ascertained. Finally, the enumeration area (the holding) is not comparable to the publication area (the parish). In many cases, holdings and parishes are very different in their distribution and attempts to portray small-scale land use change from parish data can lead to gross errors (Coppock 1960). In spite of the fact that parish data by themselves are unsatisfactory, if they are combined into ADAS districts comprising 30–40 parishes and then analysed by computer, they can reduce not only the parish boundary problem but also difficulties caused by the sheer bulk of the agricultural census data. The net result can be a description of broad land use trends, from the sub-regional to the national scales, as shown in Figure 3.3 and Table 3.13 (Coppock 1974).

From land use plans and published maps

Local planning authorities are a useful source of land use data. Their Development Plans consisted essentially of land use maps, and records of planning permissions can show the detailed pattern of change where development has occurred. Unfortunately, most of this data is confined to built-up areas, but Best has combined it with agricultural census data to produce an overall pattern of broad land use change in which he recognises two main themes:

(a) Both forest and urban land take up the same amount of land (about 9% of Britain) and have been increasing at about the same rate (1% per decade) since the 1920s;

(b) Agricultural land though declining still accounts for over three-quarters of all land (Best 1976a).

These figures have however been attacked both for overestimating and underestimating land use change. From systematic point samples of 1:10 000 and 1:2500 maps, Fordham suggested that Best's figures for the 1950s using Development Plan data could be an overestimate by as much as 400 000 ha; but Coleman has suggested that Best's figures are an underestimate (Fordham 1975, Rogers 1978, Coleman *et al.* 1974). Further

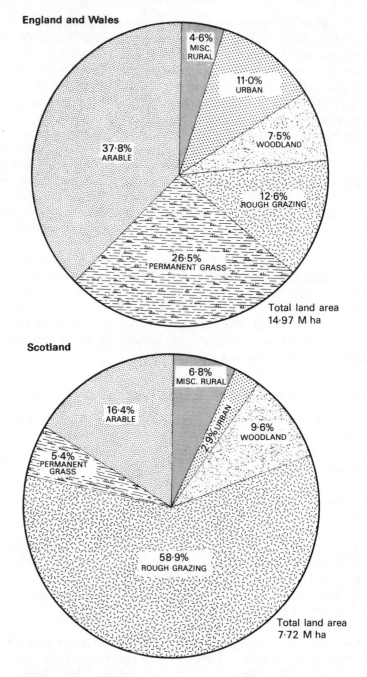

Figure 3.3 The national pattern of land use in England and Wales and Scotland. (Nature Conservancy Council 1977. *A nature conservation review: towards implementation.* London: The Council.)

Table 3.13 Changes in the major land uses of England and Wales 1900–60. (Peters, G. H. 1970. Land-use studies in Great Britain. *J. Agric. Econ.* **21,** 171–213).

Year	Agriculture		Forest		Urban		Residual	
	'000 ha	%	'000 ha	%	'000 ha	%	'000 ha	%
1900	12 570	84	770	5	810	5	883	6
1935	12 300	82	858	6	1 134	8	741	5
1960	11 920	79	1 030	7	1 618	11	465	3

work by Anderson has subsequently tended to confirm Best's original conclusions (Anderson 1977). She employed systematic point sampling on the 1:63 360 maps of the Ministry of Agriculture Land Classification Maps on which non-agricultural land uses such as urban and forest land are recorded for the mid-1960s. A sample of 3039 points produced a close fit with Best's figures as shown in Table 3.14.

Table 3.14 Structure of land use in England and Wales and EEC. (Anderson, M. A. 1977. A comparison of figures for the land-use structure of England and Wales in the 1960s. *Area* **9,** 43–5; and Rogers, A. (ed.) 1978. *Urban growth, farmland losses and planning.* Ashford: Rural Geography Study Group, Institute of British Geographers.)

	England and Wales			EEC
Land use (%)	Anderson (mid-1960s)	Best (1961)	Best (1971)	Rogers (1971)
agriculture	79·1	79·1	76·9	64·2
forest	6·9	6·9	7·5	21·6
urban	9·7	9·9	11·0	6·8
miscellaneous	4·2	4·1	4·6	7·4

Concern about the lack of reliable land use change data and the differences revealed by the estimating techniques outlined above, encouraged the DOE to initiate a number of experiments and innovations in the area of land use data collection, but by 1978 none had produced a workable technique. The most promising was the 1974 instruction to local authorities to return records of land use change, involving development, to the DOE for annual collation (DOE 1974e). Even this, at first sight easy instruction, has run into major difficulties because of the sheer size of the data base and uncertainty about how to classify multiple land uses. To overcome these problems the DOE has issued three volumes demonstrating how land use data may be collated, classified and processed:

The General Information System for Planning (GISP) (DOE 1971a);
National Land Use Classification (NLUC) (DOE 1976b);
Spatial retrieval for point referenced data (DOE 1976e).

None has yet overcome the apparently insoluble problems of land use data collection, most notably:

(a) Whether land use should be recorded by form or by function (appearance or activity)?
(b) What should be the spatial unit? Grid references, post codes, Basic Spatial Units (Rateable hereditament units and fields) or activity curtilages.
(c) How can the collection of a total data set be made less time consuming and inefficient compared to sampling? (Dickinson & Shaw 1978, Coppock & Gebbett 1978.)

In conclusion, local authority sources and published maps (primarily the Ordnance Survey series) allow broad statements of land use change but for a number of reasons, notably irregular resurvey and variable definitions, they cannot form a nationally consistent or locally accurate data base for measuring land use change. Indeed, a 1978 report by the Experimental Cartography Unit drew attention to the inadequacies of existing data (including that derived from development control records) as a data base for either monitoring current land use changes or evaluating any future changes which may arise from the implementation of structure plan priorities (Experimental Cartography Unit 1978).

From aerial photographs

On first sight, the development of air photographs over 50 years ago would seem to have answered many of the problems of land use data collection, for they can be consulted at will and can easily be copied and translated into map form with or without the use of expensive or sophisticated manpower and machinery. However, air photographs have not yet taken over from field survey as the most cost-effective way of producing data on land use change.

Surveys are very costly to fly, and even purchases of existing cover can be prohibitively expensive. Air photographs also only represent one point in time, which dates them very quickly, in comparison to Ordnance Survey maps that are revised for major changes by local offices of the survey and open to inspection. Perhaps most significant of all is how to interpret the unfamiliar overhead view presented by aerial photographs and the fact that the most economical scales, 1:10 000 and above, make many land uses indistinguishable from each other. In the future, scanning techniques may become sufficiently accurate to allow the continuous records received from orbiting satellites to be interpreted, but at present the accuracy rate of about 80% is unacceptable (Barret & Curtis 1974, Goodier 1971).

In the meantime, aerial photographs have yielded some useful data, particularly about urban land. In 1975 the DOE decided to map all urban land (developed land) covering at least 5 ha from 1969 black and white RAF photographs at the scale of 1:60 000 (DOE 1978b). Using a digitiser, a national data bank of urban land was built up, subdivided into the five categories of residential, industrial/commercial, educational, transport and open space, and maps of these land uses have been published at the 1:50 000 scale. Of most interest is the figure for developed land, which at 9·8% of England and Wales in 1969 closely confirms Anderson's total of 9·7% for the mid-1960s shown in Table 3.14.

From land utilisation surveys

There have been two nationally comprehensive land utilisation surveys, the first conducted by Sir Dudley Stamp in the 1930s, the second by Alice Coleman in the 1960s and 1970s. The 1930s survey had several advantages over the more recent survey for measuring rural land use change. It was rurally based and concentrated on agricultural land use; each county was the subject of a major piece of research work and each set of County maps was accompanied by a well-researched County volume (Stamp 1962); the survey was completed in only two to three years and so provided a consistent pattern for the whole country.

There were however a number of disadvantages, especially the small scale of publication, 1:63 360. This was particularly unfortunate since the field survey was at a scale of 1:10 560 and many of the original maps were lost in the war. The 1930s were also a very depressed period for agriculture and so the maps showed agriculture in its most run down state since the Middle Ages. Furthermore, the base maps were often two or three decades out of date. Nonetheless, the survey and its accompanying County volume did demonstrate the loss of agricultural land taking place, the run down state of agriculture, and the under-recording of agricultural land by the census, and so did much to hasten the introduction of the post-war land use and agricultural planning system.

The Second Land Utilisation Survey was conducted rather irregularly throughout the 1960s and 1970s, and while some areas have been surveyed twice, other areas were not first surveyed until several years after the inauguration of the survey in 1961 (Coleman 1961). It used the same 1:10 560/1:10 000 scale of its predecessor but doubled the number of categories of land use to well over 60. Publication has been at the more effective scale of 1:25 000 (not available in the 1930s) and these published maps show 52 categories of land use. Like the First Survey, the work relied almost entirely on volunteer labour although paid staff have been employed at certain times. The Second Survey has highlighted the continuing process of land use change, both within and outside the orbit of planning control, and Coleman has used the Survey to argue that planning has had little effect on slowing down the rate of land loss to urban growth. It has also allowed Coleman to introduce the useful concept of 'scape' types based on land use configurations: townscape, urban fringe, farmscape, marginal fringe and wildscape, providing a useful approximation of the five major landscape types present in Britain (Coleman 1969). Much of the utility of the Survey has, however, been nullified by the sporadic progress of the actual field survey, which means that adjacent sheets are not always contemporaneous. The accuracy of some of the sheets is also questionable, a problem encountered with all volunteer surveys.

Table 3.15 Use of agricultural land in Britain 1939–75. (Central Office of Information 1967, 1977. *Agriculture in Britain*. London: HMSO.)

FIG 1.

'000 ha	1939	1965	1975
wheat	720	1 026	1 035
barley	404	2 183	2 345
oats	971	410	233
potatoes	284	300	204
sugar beet	74	184	198
temporary grass	1 660	2 660	2 138
permanent grass	7 611	4 912	5 074
rough grazing	6 680	7 215	6 555
'000			
beef and dairy cattle	3 900	5 000	6 044
sheep and lambs	25 800	29 900	28 270
pigs	4 500	8 000	7 532
poultry	76 200	118 200	136 610

Nonetheless, the two Surveys provide an invaluable starting point for a study of land use change although there are problems of comparison. They are drawn on different map projections and they do not always match up and the land use classifications are not always directly comparable (Coleman, Isbell & Sinclair 1974), although Coleman has partially circumvented these problems by redrawing and reclassifying the Stamp data onto a modern base. A fundamental weakness of both surveys is that they are limited to land use data only and it is to be hoped that future surveys will include information on habitat type, age of trees, and landscape value, among other things, to make them more useful to rural land use managers.

Summary

In combination the sources of data outlined above and particularly the agricultural census allow a broad portrait of post-war agricultural change to be drawn up as shown in Table 3.15. The main feature of the table is the increase in the area devoted to arable crops, notably, barley, wheat and sugar beet. The decline in two crops, oats and potatoes, is due in the case of oats to the virtual disappearance of the farm horse and in the case of potatoes to a combination of falling demand and better yields. Both the increased areas of arable crops and temporary grass have mainly been used to feed the greater numbers of livestock, especially poultry, pigs and cattle.

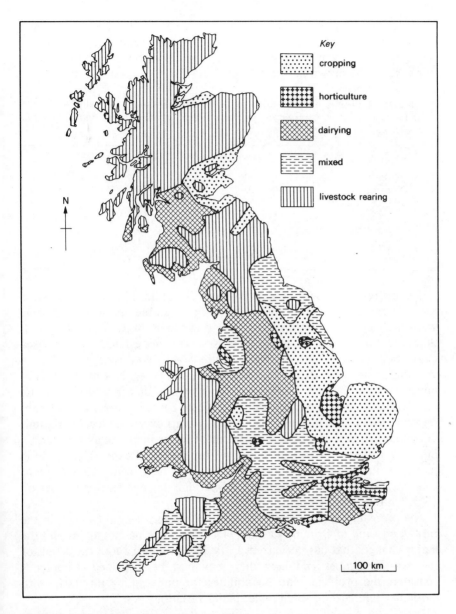

Figure 3.4 Farming regions of Britain. The broad patterns are dictated by physical conditions of relief, slope, climate and soil, but sub-regional variations reflect socio-economic factors like market price, government policy and tradition. (Gilg, A. 1978. *Countryside planning*. Newton Abbot: David and Charles.)

The area of permanent pasture and rough grazings has declined as new husbandry techniques allow arable production or temporary grass to take over. Agricultural production has also become more integrated between regions, with livestock in the pastoral west relying on imports of food and fodder from the arable east, and this movement of produce has encouraged the tendency for farms to become more specialised and has led to a loss of mixed farms as shown in Table 3.16.

Table 3.16 Full-time holdings by type of holding. (Edwards, A. and A. Rogers (eds) 1974. *Agricultural resources: an introduction to the farming industry of the United Kingdom.* London: Faber.)

Type of farm '000	1963	no. 1970	1963	% 1970
dairy	65	50	40	38
livestock	23	21	15	16
pigs/poultry	9	9	5	7
cropping	25	24	16	18
horticulture	16	13	10	10
mixed	23	13	14	10
total	160	130	—	—

The net result of these changes has been four-fold. Firstly, the increased area of arable land has meant an extension of arable landscapes not only westwards but also upwards into the downlands of southern England. Secondly, the increase of temporary grass has further extended arable based cultivation systems into formerly more stable and ecologically more valuable areas of rough grazings and permanent grass. Thirdly, the loss of mixed farms in the central area of the country has further reduced land use and landscape diversity. Fourthly, there has been a general loss of sub-regional diversity as modern production systems overcome local variations in soil and climate to produce a pattern of farms increasingly specialised, intensive and stereotyped in their land use and landscape. The resulting pattern is illustrated in Figure 3.4, and shows the growing dominance of four basic farming types: cropping, mixed, dairying and rearing (Anderson 1975, Church 1968, Coppock 1971).

The preceding analysis has examined the reasons behind changes in the rural landscape of Britain, ways in which these can be measured, and the major changes that have occurred. Little has been said about the details of the new landscapes and how they may best be described. Chapter 4 considers the problems and possibilities for undertaking this task, with particular reference to small-scale spatial variations.

4 *Planning and land use*

It should be clear from the preceding chapters that the contemporary countryside reflects the complex interaction of many different factors. The resulting changes have been so rapid and dramatic that both government action and new attitudes are necessary if the environmental resources of Britain's small and crowded landscape are to stand any chance of being permanently conserved. Given the regional variations in the overall pattern of land use, it is unreasonable to expect that any single remedy will be adequate to cope with all the new developments now occurring in rural areas. Nevertheless, the changes may be broadly grouped into four, each corresponding to a major landscape type: the urban fringe, the agricultural lowlands, the hills, and the uplands. These are roughly equivalent to the rural components of Alice Coleman's landscape classification (Coleman 1976). Each of the four is now considered in turn with reference to detailed case studies, to see how the landscape of each has altered and how far planning policies and powers have been able to manipulate and control the new order.

The urban fringe

The fringes of urban areas have always been subject to rapid change. They are zones of transition from urban to rural land uses and the change continues unabated, in spite of development control powers and the sanction of Green Belt designation, which has been widely copied following its initial success around London in the post-war era. It is, however, particularly difficult to measure accurately land use changes in these areas because of the enormous diversity of land use and the speed and subtlety of many of the new developments. It may be argued that the urban fringe is more 'urban' than 'rural' but, since the whole purpose of Green Belts, for example, is to check the rapacious quest for building land and as they cover over 11% of England and Wales, it is unthinkable that they or any other part of the urban fringe be omitted from a comprehensive assessment of rural land use change in the past 30 years.

A study of London's urban fringe in the early 1960s overcame some of the problems of measurement by combining aerial photograph data with ground surveys. It recorded 11 categories of land use at a scale of 1:25 000 for the whole survey area, most of the categories being further subdivided to give a more precise picture where necessary. The overall distribution of land use was obtained from a systematic line traverse of the area; changes in use between 1955 and 1960 were ascertained by comparing aerial photographs with the 1:25 000 maps. The problem of how to record multiple land use was not overcome completely and a partial solution involving the apportioning of multiple use equally between the relevant categories had to be employed. Nevertheless, the study found some interesting patterns: only 11% of the Green Belt was used for activities not allocated in the appropriate development plans; 70% of the land was still in agriculture and only 5% and 6% respectively were used for recreation and residential purposes (Thomas 1964).

A later study of part of the same Green Belt in the Slough and Hillingdon areas of London did not confirm the apparently satisfactory state of affairs (MAFF 1975, Low 1973). Using Ministry of Agriculture census data, it concluded that, if the losses of agricultural land to other uses continued at the present rate, there would be no farming at all by 1991. It is possible, however, that the conclusions of this latter study were unduly pessimistic, since the period of study included permissions for two major but very atypical developments, Brunel University and an extension to Heathrow Airport's runway.

The run down and transference of farmland to other uses like recreation and institutional purposes has also been studied by Rettig in the Tyneside urban fringe (Rettig 1976). He compared aerial photographs with the Second Land Utilisation Survey, and found an increase in scrubland at the expense of farmland, a trend subsequently noted elsewhere by Coleman (Coleman 1978). The Tyneside study also included the results of an interview survey of farmers in the area, which initially asked questions on tenure, length of stay, and type of farming, as well as eliciting reactions and attitudes to farming in the Green Belt. Surprisingly, it revealed that in spite of fragmentation of farms due to motorways (24%) and state developments (16%), fear of losing land to urban growth (37%) and problems of trespass (87%), nearly three-quarters (72%) of farmers would still choose to farm in the Green Belt, even if given the opportunity to farm elsewhere.

Another study, this time of the Oxford Green Belt, also identified many non-conforming uses and argued that the policies had only been partially successful (Rural Planning Services 1979). All the statutory undertakers and in particular the electricity industry were singled out for criticism. The general conclusion was that the key to environmental health and preservation was a strong and well-protected agricultural industry and the main recommendation was a comprehensive package of measures to encourage sound farming practice, as well as proposals for improving the visual environment, removing detractors, providing more recreational facilities, and controlling such further urban development as may be necessary.

It is difficult to generalise in too much detail about the whole variety of urban fringe areas where Green Belts have been designated, but Hebbert and Gault have attempted to summarise all the policies in the 27 counties and 140 districts affected (Hebbert & Gault 1978). They found that there has recently been a tendency for planning authorities to broaden the rationale for Green Belts, largely because of stronger demands for recreation and because farmers are less willing to act as unpaid park keepers. The authorities are having to face up to the problem of how to do something worthwhile with land preserved from development by Green Belt restrictions. In other words there is a growing conflict between negative control and positive management in the urban fringe (Davidson 1976). This is discussed more fully in Chapter 7, through an assessment of the Countryside Commission's Urban Fringe Experiment (UFEX).

By no means all the urban fringe is designated as Green Belt and many of the areas under the severest pressure are only subject to normal planning controls. A study conducted by the authors in east Devon looked at the changes that had occurred on a strip of land between Exeter and the coast. The main features of the area are a band of high quality farmland to the immediate south-east of Exeter's urban fringe, a broad belt of heathland and woodland in the centre, and the coastal valley of the river Otter beyond this. The whole area is under extreme development pressure of a number of different kinds, needing to accommodate the residential expansion of Exeter, Exmouth and Budleigh Salterton, the commercial development stimulated by the arrival of the M5 motorway, and the recreational use of the coast and the Exe estuary. At the same time much of the land is either of Grade 1 or Grade 2 agricultural value and nearly a half of the area is included within an AONB. It thus provides a classic instance of conflicting land use pressures and a lack of effective policies for resolving the competing claims. This is underlined by Figure 4.1, which reveals two fundamental features common to most rural land use planning policies: the white land syndrome, where large areas are left without any policy guidance, save a presumption against new development; and the concentration of policies into certain 'key areas', producing conflicts of interest where two and sometimes three policies cover the same piece of land. In east Devon not only do nearly all the development policies occur within the AONB, they are also concentrated along the heathland ridge and the coastline.

Figure 4.2 illustrates the land use pattern in more detail, showing in particular the greater amount of developed land to the north-west, on the fringes of Exeter. Table 4.1 shows the same information in tabular and percentage form, revealing that aleady less than two-thirds of the area is in farming, and that the high figure of 12·4% is given over to developed land, the latter figure rising to 13·3% if mining and industry are included.

Figure 4.3 shows the overall pattern of land use change and highlights a number of interesting points. Any large amounts of land use change are

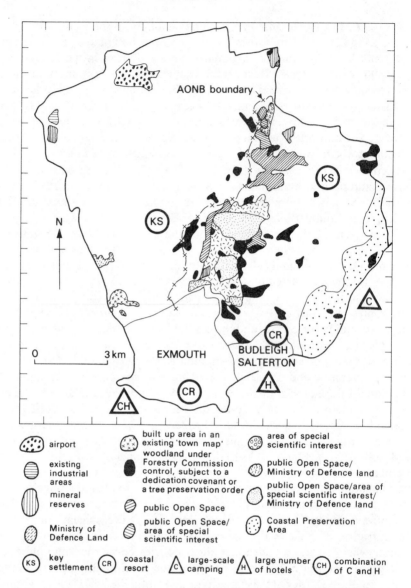

Figure 4.1 Planning constraints in the east Devon study area, recorded on the Devon Development Plan, second review. (Devon County Planning Department 1972. *County of Devon Development Plan*. Exeter: The Department.)

confined to a few areas, only 20 out of a total of 165 kilometre squares having changes in excess of 10 ha (10%) and 116 squares having a change of less than 5 ha (5%). What change there is, is focussed at a few key locations: the urban fringe of Exeter with the motorway and the new industrial estate, the heathland ridge, and the belt of land running east–west across the

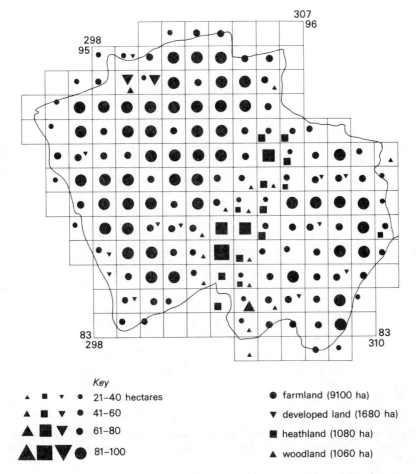

Key

▲	■	▼	●	21–40 hectares
▲	■	▼	●	41–60
▲	■	▼	●	61–80
▲	■	▼	●	81–100

● farmland (9100 ha)

▼ developed land (1680 ha)

■ heathland (1080 ha)

▲ woodland (1060 ha)

Figure 4.2 Major land uses in the east Devon study area 1973. (Authors' survey.)

northern half of the area where there has been a significant loss of orchards. It is worth noting that the first two of these areas coincide with the policy areas shown in Figure 4.1, suggesting that planning policies have picked out

Table 4.1 Land use in the east Devon study area. (Authors' survey.)

Land use	ha	%	Land use	ha	%
farmland	9100	67·3	other land	240	1·8
developed land	1680	12·4	orchards	190	1·4
heathland	1080	8·0	mining	70	0·5
woodland	1060	7·8	industry	60	0·4
			water	50	0·4

total area 13 530 ha

the main areas of change even if they have not at first sight been very effective in determining its nature and direction.

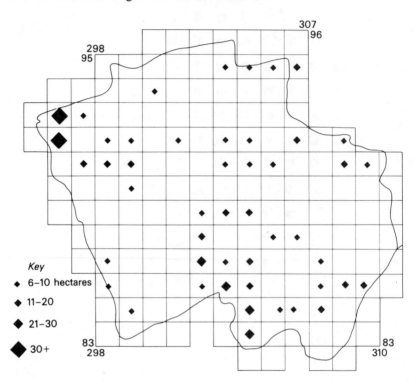

Figure 4.3 Land use changes in the east Devon study area. Most change has occurred at the junction of heathland and farmland, and around Exeter's urban fringe. (Authors' survey.)

A more detailed analysis can be made from Table 4.2, which shows that the four main land uses of the area also dominate the pattern of change. In the loss category, farmland and heathland account for 670 of the 845 ha lost, and orchards with 131 ha account for most of the remaining 175 ha. In the gain category, the position is less clear with farmland, developed land and unused land (often land in the process of development) accounting for only 679 of the 845 ha involved. The biggest single gain was recorded by woodland, but if unused land, industrial land and developed land are taken together they become the single largest category with 291 ha. In terms of net losses/gains, farmland, heathland and orchards recorded large net losses, while woodland, developed land and unused land recorded large net gains. In total there was a change of use in no less than 6·24% of the area, a surprisingly high figure. A systematic way of assessing the impact of planning powers is to subdivide the largest changes according to the degree of influence that planning can exert on them as shown on page 60:

Table 4.2 Land use change in the east Devon study area c. 1950–74 (ha). (Authors' survey.)

From/to	Orchard	Farm-land	Wood-land	Devel-oped land	Park-land	Heath-land	Water	Unused land	Mining	Industry	Total loss	% of area loss/gain	Net area loss/gain
orchard		114	10	7							131	0·97	− 113
farmland	18		95	129	9	54		88	16	39	448	3·31	− 254
woodland		12		2				12		3	29	0·21	+ 202
developed land*								8			8	0·06	+ 137
parkland		3						1			4	0·03	+ 5
heathland		65	126	7					22		222	1·64	− 168
water						2							+ 2
unused land													+ 109
mining													+ 35
industry										3	3	0·02	+ 45
total gain	18	194	231	145	9	54	2	109	38	45	845		
% of area	0·13	1·43	1·71	1·07	0·07	0·40	0·01	0·80	0·28	0·33	6·24	6·24	

total area 13 530 ha

*Does not include mining or industry.

Direct (compulsory) negative powers over change

farmland to developed land	129 ha
farmland to industrial land	39 ha
heathland to mining	22 ha
farmland to mining	16 ha
total	206 ha

Indirect (voluntary) positive powers over change

heathland to woodland	126 ha
orchard to farmland	114 ha
farmland to woodland	95 ha
heathland to farmland	65 ha
total	400 ha

No effective powers over change

farmland to unused land	88 ha
farmland to heathland	54 ha
total	142 ha

It is clear that the most widespread changes have occurred in those categories covered by indirect positive powers, in particular the improvement grants and subsidies paid under the Agricultural Acts and the CAP, and the tree planting grants paid under the Forestry Acts. Within this group the biggest change, heath to woodland, occurred in only 20 squares along the heathland ridge and was the result of both scrub reversion and intentional planting. Planning policies have directly affected the process by encouraging planting with grant aid and by allowing access and military training, thus reducing the agricultural potential of an already poor area to virtually zero and leaving the heath open to woodland invasion. The change from orchards to farmland is only slightly smaller, but is far more widespread, occurring in nearly 60 squares. It reflects the decline in farmhouse cider making, causing orchards to be grubbed up to make way for intensive grazing for dairy cows. This has also been encouraged by farm improvement grants. The conversion of farmland to woodland occurs widely throughout the area (in about 40 squares) and cannot really be related to any single factor, except that it occurs mainly on marginal land. Accordingly, it is fair to assume that the causes are as much accidental as the result of planting grants, and therefore augurs well for the strategy advocated by the *New agricultural landscapes* study (Ch. 7). Finally, the change from heath to farmland is concentrated along the marginal fringe of the heathland ridge and once again is most probably related to improvement grants, for it mostly occurs in large, newly enclosed fields. However, the total area involved is nearly counterbalanced by the reversion of farmland to heathland and emphasises the need to provide data on two-way flows in land use studies as has been done in Table 4.2. In summary, planning policies have hastened the changes outlined above but have by no means been the only, or even the prime factor involved.

The next largest group of changes occurred in those categories covered by direct negative powers, by far the most important of which were the development control powers exercised by the local planning authority. Largest and most widespread was the loss of farmland to developed land that occurred in over 60 of the 165 squares, and was contrary to the policy of containment outlined in Figure 4.1. Conversely, the loss of farmland to industry is highly concentrated in a newly provided industrial estate in the extreme north-west of the area, showing that not only can planning powers contain industrial growth, they can also steer it to selected locations. In the last two sets of changes, heathland to mining and farmland to mining, the losses again conflict with other planning policies. The heathland ridge is designated in the Development Plan as land for open space and nature conservation, but neither these policies or designation as an AONB seems to have had much effect on the changes, indeed, as Figure 4.1 shows, part of the AONB is actually allocated for mining on the Plan.

The final group of changes, and the one over which planners have no real power, is also the least significant, especially since most of the 88 ha recorded as land lost from farmland to unused land, are due to motorway construction, and somewhat oddly classified by the Coventry-Solihull-Warwickshire landscape evaluation method employed by the authors. Nonetheless, small pockets of change to unused land occur in nearly 40 squares, giving credence to both Alice Coleman's thesis that planning has not prevented the creation of many pockets of unused land in the urban fringe, and to the contention of the *New agricultural landscapes* study, that much wildland can still be retained in a prosperous farmed countryside. The final change from farmland to heathland, reflects a reversion of marginal fields to heath around the fringe of the heathland ridge.

East Devon is an urban fringe landscape subject to a great deal of development pressure, and this study has shown that a considerable amount of complex and often contradictory change is taking place there. Most of it has occurred in the zones of transition between one dominant land use and another and also corresponds with the main centres of interest of planning policies. About 85% of the changes are covered by planning policies in some degree, but detailed investigation has revealed that these policies have only exercised effective control in about half of these. Planning cannot be said to have had a fundamental impact, except in the immediate hinterland of Exeter where decisions have tended to encourage rather than limit the growth of the city even though much of the land is of Grade 1 agricultural quality. On the heathland ridge, where change has been greatest, most of the developments have been outside the direct scope of planning. It is possible, however, that this part of east Devon would have altered even more radically but for planning controls: two championship-standard golf courses and all their ancillary facilities were refused planning permission and the greater part of the coastline has been left untouched, despite enormous potential pressure for development.

All the studies of urban fringe landscapes, including that of east Devon, have highlighted the subtlety of the land use changes occurring as agriculture gives way to urban development and greater recreational and institutional use. All argue that the best way to preserve the landscape in such areas is through a multiple land use strategy, but in general there is a paucity of ideas as to how the various interests can be reconciled side by side in the same area.

The agricultural lowland

There have been a number of studies indicating the general rate and pattern of land use change in the agricultural lowlands but, until recently, there was no satisfactory explanation either as to why and how these changes were taking place, or their precise impact on the landscape, or the effect that planning policies have had. For example, studies like Brett's work on the Thames valley were mainly subjective and emotive in their appeal, while review articles relied heavily on secondary sources (Brett 1965). They did however pick out the main trends, namely radical changes in the physical layout and production systems of farms leading to a more uniform and open landscape (Leonard & Cobham 1977, Davidson & Lloyd 1978). The traditional, intimate and enclosed landscape, closely related to local variations in geology, soil, slope and drainage, depended on two pre-conditions which have now been removed. First, the need to be reasonably self-sufficient in farm supplies, to recycle waste products and to conserve the farm's basic resources by conservative husbandry along the lines of the Norfolk four-course rotation. Secondly, a surplus of cheap farm labour, which could be employed during seasonal lulls in productive work to maintain the detailed fabric of the small-scale farm landscape by exercising such skills as hand-hedging and ditching. Now that modern technology has removed the need for farms to be self-sufficient and has pared the labour force to the bare minimum, the intimate, enclosed farming landscape can no longer be maintained. The post-war period has seen a switch from a mixed, enclosed, handmade landscape, to a simple, open, machine-made landscape. In addition the increased use of liming, drainage and chemicals has fundamentally altered the detailed ecosystem of rural areas on which even large features of the landscape like trees still ultimately depend (Green 1976).

In order to examine these changes in more detail a number of studies of farming change have been conducted in the 1970s, the most comprehensive being the study of *New agricultural landscapes* commissioned by the Countryside Commission, which had the following brief, 'To find out how agricultural improvement can be carried out efficiently but in such a way as to create new landscapes no less interesting than those destroyed in the process' (Countryside Commission 1974, p. 1).

The study did not seek to reverse or halt the process of agricultural change, but rather sought to see if it could be adapted to meet the wider demands being put on the countryside. Essential to this approach are the farmer's motives, his attitudes to and perception of change, and the way in which each individual farmer views the various features in the landscape. For example, depending on one's point of view, a ditch can be either a drainage line, a source of irrigation, a wildlife habitat, a source of pests, or a fish reserve and so on. The study thus concentrated on the attitudes of individual farmers to change in seven fairly small farming areas thought to represent a reasonable cross section of farming types: Cambridgeshire (intensive arable), Huntingdonshire (extensive arable), Dorset (extensive arable), Somerset (dairying), Herefordshire (mixed), Yorkshire (general cropping) and Warwickshire (livestock, rearing, dairying and mixed). In each area an examination was made of the motives for change and its scale and impact.

In each study area seven or eight farmers were interviewed to ascertain:

(a) Size, tenure, farm type, land use,
(b) Landscape change since 1945, both intentional and accidental,
(c) Main reasons for change,
(d) Work involved in maintaining landscape features,
(e) Attitudes towards game,
(f) Attitudes to outside factors, like planning controls and recreation,
(g) Attitudes towards conservation and its enforcement.

In addition to the farm interviews, there were a number of field surveys, including a visual survey of the horizon from ten randomly selected roadsides; a survey of hedgerow and park trees by species, age and by type of boundary; a hedgerow survey by quality of wildlife habitat and type of boundary; a survey of wildlife habitats including trees, hedges, wetlands, ponds, rivers and buildings; and an analysis of land use change based on a comparison of the First and Second Land Utilisation Surveys with updating of the Second Survey where necessary.

As Table 4.3 shows, the most dramatic developments have been the enlargement of field size and the loss of hedgerow trees. In addition, many other changes of a more subtle but equally fundamental nature were found. Streams, ditches and rivers had been cleaned and straightened, permanent grassland with its rich habitat had been ploughed up, and modern farm buildings, often out of scale and mass produced, had replaced small buildings constructed from local materials.

Most of these changes were forced on farmers by economic necessity and only those with substantial incomes could consider the retention of features likely to impede maximum agricultural production. Nevertheless, personal preference has also often been of considerable importance and farmers have been prepared to forgo immediate economic gains if they had some interest, such as shooting or hunting, in preserving the landscape. Not surprisingly they were not generally in favour of negative powers like development

Table 4.3 Changes in field size and hedgerows 1945–72. (Countryside Commission 1974. *New aricultural landscapes.* Cheltenham: The Commission.)

	Cambridgeshire	Huntingdonshire	Dorset	Somerset	Herefordshire	Yorkshire	Warwickshire
average field size 1945 (ha)	5·7	7·7	7·3	3·6	4·5	6·1	4·9
average field size 1972 (ha)	13·0	18·2	8·5	5·3	6·5	7·7	6·1
increase in field size 1945–72 (%)	128·0	137·0	16·0	44·0	45·0	26·0	25·0
length of hedges removed per ha (m)	35·6	28·0	6·0	15·1	14·3	11·4	9·8
length of hedges remaining per ha (m)	—	44·0	63·6	94·7	149·2	71·2	125·8
number of hedgerow trees per 100 ha							
1947	15·8	23·9	6·9	20·2	19·8	20·6	27·9*
1972	2·0	4·9	2·8	6·1	16·2	13·4	32·8
1972 as % of 1947	13·0	20·0	41·0	30·0	82·0	65·0	117·0
age distribution 1972							
mature	11	31	39	30	34	50	22
semi-mature	60	31	16	60	37	34	50
sapling	23	18	43	9	27	5	23
dead	6	20	2	1	2	11	5

*1897 not 1947.

control, or Tree Preservation Orders, without financial compensation, or incentives for positive works.

Using this as evidence, the report of the *New agricultural landscapes* initiative proposed a flexible attitude to land use changes, except in those few areas of outstanding importance where it recommended that any change should be resisted. It recognised the inevitability of new developments, but pointed out that there was still a basic subdivision into productive and non-productive land. The report suggested that on land which was naturally unproductive due to steep slopes, wetness or stream banks, or artificially unproductive due to some obstruction like ownership boundaries, roadsides and awkward corners, semi-natural vegetation should be retained and if possible encouraged. Since unproductive land tends to be linear, this would provide a visual continuity and a wildlife communication network through the agriculturally productive landscape. The authors acknowledged that the net result would be a very different landscape, but argued that it need not be a worse one and could be better.

The ideas were broadly endorsed by the Commission and in 1977 they put forward five main objectives for future policy:

(a) Stop unnecessary change,
(b) Ensure advice and aid is available,
(c) Ensure that public agencies set a good example,
(d) Develop higher standards of development control,
(e) Attract and maintain the interest of education and the mass media (Countryside Commission 1977).

Although *New agricultural landscapes* is perhaps rather limited and is certainly more sympathetic to economic farming than conservationists would like, it represents an important step forward. As Chapter 7 shows, its proposals are now being tried elsewhere.

Similar studies in Hereford & Worcester have endorsed the findings and its conclusion that the key to successful management of the landscape is an enlightened attitude on the part of farmers to multiple land use, particularly on unproductive land (Hereford & Worcester 1976). A number of pioneering studies and experiments have also shown that farmers are willing to co-operate in management schemes and that profit margins do not normally suffer as a consequence. The Countryside Commission is eager to encourage them and has recently started the 'Demonstration farm project' which manages selected holdings according to progressive conservation principles. It is too early to say how successful the experiment will be but the omens are good as it is based on established, independent schemes, such as that mounted by Essex County Council, the ICI demonstration farm at Crewkerne and the Prestwold estate in Leicestershire (Essex County Council 1976, Hart 1977, Whitlock 1975). Central to all these initiatives is a strategy of retaining wild corners and planting up less productive land alongside linear landscape features, thus providing proof that farming is compatible with wildlife and landscape conservation.

Table 4.4 Percentage of farm area retained as wildlife habitat. (MAFF 1976. *Wildlife conservation in semi-natural habitats on farms*. London: HMSO.)

	Arable	Mixed	Dairy	All farms
less than 5	64	51	17	35
5 to 15	33	46	80	56
15 to 30	1	3	3	5
over 30	2	0	0	5

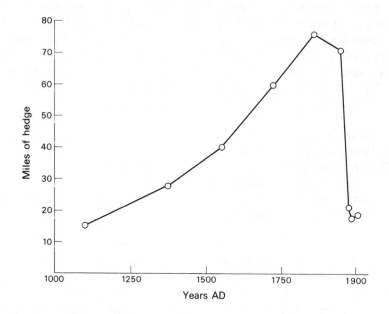

Figure 4.4 Length of hedgerow on 1800 ha in Cambridgeshire 1000–1970. The post-war period has seen the elimination of over 800 years of hedgerow history. (Hooper, M. 1977. Hedgerows and small woodlands. In *Conservation and agriculture*, J. Davidson and R. Lloyd (eds.). Chichester: John Wiley.)

The attitude of the farming community in general to those new ideas has yet to emerge fully, but the reaction will be vital, for it is not only farmers' individual actions on their own farms that matter. A survey of East Anglian farmers showed that most rural planning agencies are dominated by farming representatives, either as elected councillors or as NFU nominees (Newby *et al.* 1977). Indeed Newby is highly critical of the farming lobby and their stranglehold over the countryside and would probably view the official response to *New agricultural landscapes* as a cosmetic concession rather than a fundamental change of heart. A study of farmers' attitudes to conservation in the mid-1970s did in fact show that the new ideas were by no means wholly welcome. It revealed that 13% of farmers intended to remove wildlife habitats, although by the same token 13% also intended to improve existing habitats or create new ones (MAFF 1976b). There was a tendency for the larger farmers to react more positively, suggesting that the total area of wildlife habitat might increase. The overall pattern, however, is further confused by the variation in habitat percentages depending on the size and type of farm involved (Table 4.4).

Hedges and hedgerow trees are one of the most important components of wildlife habitats, but their nature and distribution have been in a constant state of flux (Fig. 4.4) and the changes have been particularly rapid since the end of the Second World War (Table 4.3). (Pollard 1974). Charting the detail of alterations to the pattern of hedges and hedgerow trees is far from easy because they are not usually differentiated from other sorts of field boundary on Ordnance Survey maps. Accordingly, the Forestry Commission has made random counts of the distribution of these features from ground surveys and aerial photographs and has estimated that the 1 million km of hedgerow existing in 1945 had been reduced at an annual rate of 6000 km a year between 1945 and 1970. A survey conducted between 1979 and 1981 will update these figures by 1983. These national figures, though useful, nonetheless conceal substantial regional variations and so in common with studies of agricultural land use change, a number of detailed local surveys have had to be undertaken.

One of these was the *New agricultural landscapes* study referred to before and, as Table 4.3 shows, the most significant losses have been in the arable eastern counties. The study found that the reasons for hedgerow removal, in order of importance, were to facilitate mechanisation and intensify production, to improve grazing management, to reduce the burden of maintenance, to remove untidy and agriculturally redundant features, to facilitate drainage, and, finally, to gain land. Conversely, the reasons given for retaining hedges, again in order of importance, were to maintain stockproof barriers, to provide shelter for stock and crops and prevent erosion, to provide cover for game, to maintain amenity features and wildlife habitats, and, finally, because there was no reason to remove them. Indeed, many of the original reasons for removal have now diminished in importance and it could be that hedgerow removal has passed its zenith.

Not surprisingly, most studies of hedgerow removal have been conducted in the arable east where the change has been greatest. A 1976 study in Norfolk by the planning department followed up in further detail, two earlier surveys, one based on aerial photographs (Baird & Tarrant 1972), the other on a Forestry Commission field survey (Penistan 1972). The planning department divided the country into seven 'landscape types' according to farmland, tree cover and geology. 200 plots measuring 500 m by 500 m were selected on a sample basis and were surveyed for all trees over 6 m in height, subdivided by species and age, and the results were recorded on 1:2500 maps, as shown on Figure 4.5. The results showed that not only had

1/2500 Field Survey Map.

Landscape type **2** Map Ref. **XY 0/** Proportion of countable Land **100** %
No. of groups **/** Length of Hedge **0·78** km. Length of hedgeless boundary **0·7** km.
Length of boundary removed **0·81** km.

	OAK	ASH	ELM	Alder					TOTAL
Semi-mature	/			8					9
Mature, healthy	/			/					2
Mature, 10–25% deadwood				/					/
Mature, 25–50% deadwood									
Mature, 50–75% deadwood	2								2
Mature, 75% + deadwood	/		ĩ						2
TOTAL	5		/	10					/6

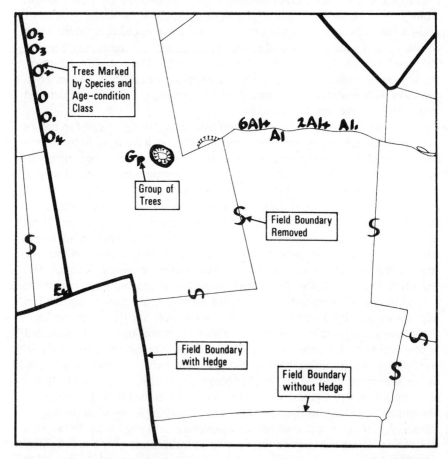

Figure 4.5 Example of a method for surveying farmland trees by age, location and species. Each species is recorded *in situ* by using a code for its age. (Norfolk County Council 1977. *Farmland tree survey of Norfolk.* Norwich: The Council.)

hedgerows and hedgerow trees declined rapidly, but that they would continue to decline because of the senile age structure of much of the remaining tree cover (Norfolk 1977). Accordingly, the study recommended a planting rate of 40 000 trees a year, merely to halt the decline. Although Countryside Commission grants could provide funds for some of this, the study emphasised the vital role that farmers could play by planting up odd corners and linear features, particularly in the areas of greatest loss.

There are many different ways of surveying trees and hedgerows. For example, work in Hertfordshire followed the technique devised by Hooper and concentrated on the composition of the hedges, notably, their floral diversity, height, width, trees present and age (Rowe *et al.* 1977). However, whatever approach is adopted, progress is slow and some form of sampling has to be employed. Possibly the most successful technique has been used by the Lake District National Park Office, which took samples of hedgerows and trees from 4 km grid squares (Lake District 1977). Trees are divided first into small woods, linear features, or point features, and then by type, condition and contribution to the landscape.

As if the removal of hedges and hedgerows were not enough, the spread of Dutch elm and other diseases throughout southern England in the early 1970s not only accelerated the change, but also extended it to those western pastoral areas that had previously been the least affected (Jones 1977, Wilkinson 1978). By 1976 as shown in Figure 4.6 many western counties, where the landscape is often dominated by elm hedges, had lost between 50 and 80% of their elms, at least half their farmland tree cover.

Fortunately, at the same time that farmland tree cover has been diminishing, the total amount of woodland has been rapidly increasing although this has not been located in the same areas and comprises different species (Table 3.13). These changes are well documented not only because the Forestry Commission own or indirectly manage (through dedication agreements) some 80% of woodland, but also because it conducts irregular surveys at roughly 10 year intervals. These demonstrate (Fig. 4.7) that coniferous species now dominate British woodland, particularly in wetter western uplands and in Scotland where coniferous trees grow best of all and can compete most effectively with agriculture. Woodland in lowland areas has fared less well and it has been argued that significant amounts of woodland in these areas are only retained because of their value for field sports, and the high cost of their clearance (Goodall 1973, Peterken & Harding 1975, Rackham 1976). Lowland woodlands are an ageing asset that will not survive without more positive management.

Further evidence of all these changes is provided by the authors' work in mid-Devon. The area surveyed covers about 10 000 ha and is made up of gentle rounded hills rising to around 150 m, and drained by the river Yeo, a tributary of north Devon's main river, the Taw. The landscape is dominated by dairying and mixed farming with the characteristic lowland pattern of hedged fields, small coppices, isolated farms and substantial villages. As

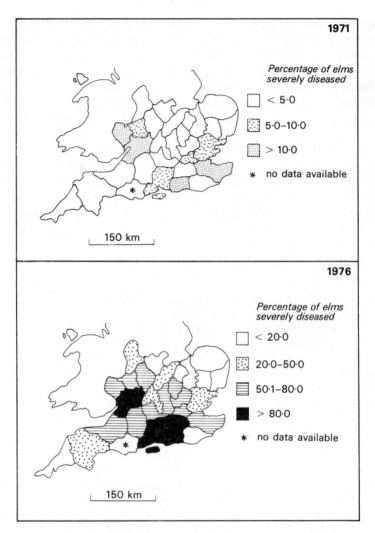

Figure 4.6 The spread of Dutch elm disease 1971–76. By 1976 most counties in southern England had lost over half their elms. (Jones, P. 1977. The spread of Dutch elm disease. *Town and Country Planning* **45**, 482–5.)

such, it is ideal for studying the lowland landscape. No one settlement dominates the area, but a number, including Bow, Lapford and Morchard Bishop, have small industries and trades dependent on the agriculture of the surrounding farmland.

Over 90% of the land is given over to farmland and Table 4.5 brings out its dominant position in the land use structure. The area devoted to urban/developed land is half the national average and only one other use, deciduous woodland (2·6%), accounts for more than 1% of total. The pattern is so regular that it did not seem worthwhile including a land use

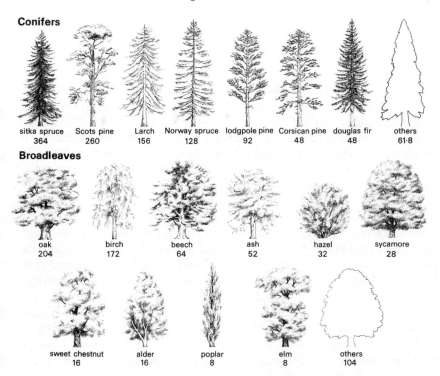

Conifers

sitka spruce	Scots pine	Larch	Norway spruce	lodgepole pine	Corsican pine	douglas fir	others
364	260	156	128	92	48	48	61·8

Broadleaves

oak	birch	beech	ash	hazel	sycamore
204	172	64	52	32	28

sweet chestnut	alder	poplar	elm	others
16	16	8	8	104

Figure 4.7 Composition of British woodland 1972. The so-called native species; oak, birch, ash, beech and Scots pine have been reduced in importance as planned planting programmes have turned to more economically suitable exotic species. (Forestry Commission 1974. *British forestry*. Edinburgh: The Commission.)

map of the area. There are in fact only twelve instances of non-farmland uses exceeding 10 ha in any one 100 ha grid square; eight cases of developed land, three cases of deciduous woodland and one of coniferous woodland.

Not surprisingly few local planning policies have been developed for the area. Indeed one of the main reasons for studying it was the lack of positive policies. Nevertheless, there are two key settlements, Lapford and Copplestone, one village of special character, Coleford, two or three small dedicated woodlands, a number of small mineral deposits, one major routeway, and a thin strip of high quality agricultural land in the south (Grades 1 and 2) surrounded by a large mass of Grade 3 land. This produces

Table 4.5   Land use in the mid-Devon study area. (Authors' survey.)

Land use	ha	%	Land use	ha	%
farmland	8889	90·1	unused land	45	0·5
developed	516	5·3	heathland	32	0·3
deciduous woodland	259	2·6	coniferous woodland	30	0·3
orchards	68	0·7	parkland and water	21	0·2
			total area 9860 ha		

a very ordinary and in theory average rural lowland area, of the type that land use planners have assumed could be left alone, except for negative development control, allowing them to concentrate their efforts where change is more rapid.

Mid-Devon has not, however, been static and Figure 4.8 reveals several peaks of quite substantial change. Furthermore, closer analysis of the source data shows that only 14, out of over 100 squares surveyed, recorded a land use change as low as 1 or 2 ha, and that there was not one complete square with no change at all. Indeed, the squares in the range 11–20 ha in Figure 4.8 are atypical peaks in a general picture of widespread *ad hoc* change. The peak areas are accounted for almost entirely by a few major gains of deciduous woodland, and a few large losses of parkland or heathland to farmland, a pattern depicted in more detail in Figure A.2.

Table 4.6 confirms not only the pace of change, which at 5·67% is only slightly less than in the more highly planned areas of the county, but also its widespread nature. There are two major flows: heathland to farmland (209 ha) and orchards to farmland (102 ha). No other transfer records a change of more than 26 ha, even though the smaller totals can sometimes conceal quite large relative changes. For example, coniferous woodland's net gain of 26 ha, raises its area from 4 to 30 ha. Three further points emerge from the table. First, there has been a gross gain of 368 ha for farmland, although losses of farmland to other uses reduces the net gain to 259 ha. Secondly,

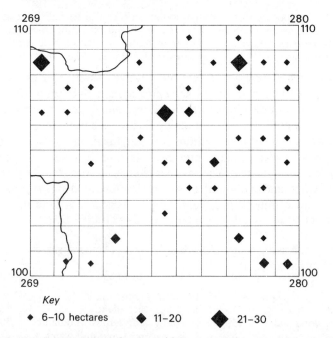

Figure 4.8 Land use changes in the mid-Devon study area. No overall pattern emerges, except for widespread small-scale change. (Authors' survey.)

Table 4.6 Land use change in the mid-Devon study area c. 1950–77 (ha). (Authors' survey.)

From/to	Orchard	Farm-land	Decidu-ous wood-land	Conifer-ous wood-land	Devel-oped land	Park-land	Heath-land	Water	Unused land	Total loss	% of area	Net loss/gain
orchard		102	11	1	9	1			3	127	1·29	−123
farmland	4		26	6	23	16	5	3	26	109	1·10	+259
deciduous woodland	21	21		22			2		3	48	0·48	+16
coniferous woodland	3	3	3							6	0·06	+26
developed land	10	10	1						1	12	0·12	+20
parkland	21	21								21	0·21	−4
heathland	209	209	23	1						233	2·36	−226
water										0	—	+3
unused land	2	2		2						4	0·04	+29
total gain	4	368	64	32	32	17	7	3	33	560	5·67	
% of area	0·04	3·73	0·64	0·32	0·32	0·17	0·07	0·03	0·33	5·67	—	

total area 9860 ha

heathland has suffered a gross loss of 233 ha and there has been so little reversion that the net loss, at 226 ha, is nearly as high and leaves a residue of only 32 ha of heathland in the entire area. Thirdly, orchards have suffered a gross loss of 127 ha and a net loss of 123 ha leaving only 68 ha of orchard in the area. In contrast to east Devon, the changes in this area are all one way, and it is quite clear that they substantiate the worst fears expressed in *New agricultural landscapes* and similar studies, that intensification of agriculture is sweeping away redundant features and unused land on lowland farms, leaving only a very small proportion of other land. In this particular case the position is alleviated somewhat by the net increase in coniferous and deciduous woodland, unused land and water, even if these increases are mostly very small. Altogether they amount to only 74 ha and only marginally offset the net loss of orchards and heathland that totals 349 ha. The net effect of these changes is a simplification of both the land use and the overall landscape pattern and it brings the spectre of a lowland countryside consisting of nothing but farmland and developed land that little bit closer.

Net transfers of land use in simple tabular form may however conceal spatial patterns more favourable to the long-term diversity of the countryside, and understate the impact of planning policies, as can be seen from the data below:

Direct (compulsory) negative powers over change

farmland to developed land	23 ha
orchards to developed land	9 ha
total	32 ha

Indirect (voluntary) positive powers over change

heathland to farmland	209 ha
orchard to farmland	102 ha
farmland to deciduous woodland	26 ha
deciduous woodland to farmland	21 ha
deciduous woodland to coniferous woodland	22 ha
heathland to deciduous woodland	23 ha
total	403 ha

No real powers over change

farmland to unused land	26 ha
parkland to farmland	21 ha
farmland to parkland	16 ha
orchard to deciduous woodland	11 ha
developed land to farmland	10 ha
total	84 ha

Direct powers over change, mainly development control, have not only kept the rate of increase in developed land at a low level in mid-Devon, they have also prevented sporadic development in the open countryside and concentrated what little development there has been into the fringes of about half a dozen existing villages, even if these have not always been the chosen key settlements. In the sense that this is an agricultural area outside the main commuting and holiday areas of Devon success is not unexpected, for the development pressures are far less than in east Devon and the other urban fringe areas discussed above.

Most land use change has occurred in those categories covered by indirect positive powers (mainly MAFF grants and subsidies and ADAS advice, which encourages land managers to intensify land use). The most significant, the loss of heathland to farmland, is the result of about a dozen large reclamations, and a number of smaller schemes, where a whole field or group of fields has been reclaimed. Grant aid and subsidy have clearly been an important factor in altering the land use in these cases. However, the small-scale reclamation that is found in over 50 squares, and usually involves only part of a field, is probably due more to an overall improvement in cultivation techniques and the more efficient use of herbicides, rather than to grant aid specifically. Nonetheless, Government policies that have encouraged technological advances in British agriculture and the price support systems which have guaranteed returns are clearly important indirect influences.

The loss of orchards to farmland is also a reflection of post-war agricultural planning policies, although the general decline in farm cider making and the poor quality of orchard grass compared to temporary leys have probably accelerated the rate at which they have been cleared. Nearly all the once numerous small cider orchards have been grubbed up, affecting well over half of the area, and markedly opening up the landscape.

In the last category, where there are no effective powers of control, the sum of a number of unrelated changes have combined to alter nearly 1% of the area, changing the landscape quite fundamentally, with no planning policies being involved at all. For example, the two-way flow between parkland and farmland represents the creation of four new parkland landscapes and the ploughing up and disappearance of an existing one. In two other cases, farmland has been allowed to revert to unused land, and orchards to woodland, demonstrating that despite the thrust of planning policies, farm production has not intensified everywhere. This is further underlined in mid-Devon, where the change from developed land to farmland reflects the rundown of farm buildings and cottages in the more remote areas and a steady move to larger and fewer farms and farmsteads.

Overall, the most striking feature of the pattern is the consistent and widespread nature of the change from heathland to farmland and from orchards to farmland. There are no isolated or quirky factors at work here; both have clearly been espoused almost universally by farmers and land managers. Indeed the loss of heathland to farmland occurs in 61 squares, and the loss of orchards to farmland in 74 squares. Other changes are less widespread, but even so farmland has been lost in 48 squares and deciduous woodland in 22 squares. The removal of nearly all redundant landscape features and the intensification of agriculture to produce a simpler and cleaner cut landscape, which characterise the pattern of change in mid-

Devon are typical of the farmed lowlands of England and Wales as a whole. Indirect positive planning powers have clearly exerted a very strong influence over the nature of the contemporary countryside. Somewhat less obviously, it would also appear that much land has only been preserved from development by the negative development control powers embodied in the Town and Country Planning Acts.

The hills

In between the farmed lowlands and the uplands there is a belt of transitional land use, which Coleman refers to as the marginal fringe and MAFF as hill land. It includes three major land uses, farmland, moorland and forestry, and it is this mixture, bounded on the one hand by lowland agriculture and on the other by upland moorland that gives these landscapes their charm and makes them amongst the most popular landscapes for recreation in Britain. They cover large parts of the country, for example in Dartmoor, central Wales, the Lake District and the Scottish Borders, but few detailed studies have been made of them. At the time of writing, however, the Countryside Commission has just sponsored an 'Upland landscape study', in the hope that it will do for the hills what *New agricultural landscapes* did for the lowlands. Most of the work previously undertaken in these areas has been as part of the background for the National Park Plans that were published in 1977.

One of the difficulties of working in this type of landscape is the problem of accessibility to such remote areas. It is not simply a lack of roads and public rights of way, but also the sheer size of the areas to be covered. To counter this the staff of the Lake District National Park organised their work into a series of complementary surveys, some covering the park as a whole, others based on a stratified random sample of 4 km^2 blocks (Lake District 1977). They concentrated on the enclosed land, since this was where change was thought to be most likely and on key features in the hill country landscapes, such as field boundaries, groups of scattered and mature trees, and the land use of enclosed fields. They found a landscape under threat, rather than one where the damage had already occurred. Of the broadleaved woodlands, 48% contained only mature and over mature trees; because these woodlands are widely used for grazing and as shelter for stock, no saplings were surviving, and they were likely to disappear from the landscape altogether if nothing were done to save them. The survey's conclusions on land use change were very tentative. Although 37% of the enclosed land had been invaded by bracken, rushes and scrub, it was not possible to show that there had been a general deterioration. Nevertheless, whatever the degree of change, it was clear that there was now considerable scope for improving productivity, especially in the large fields where the weed infestation was worst. Finally, the study also found that 38% of stone walls had either collapsed, or were completely defunct and that 66% of hedgerows were in a similar state of disrepair. It was noted somewhat wryly that although the National Park Board had no planning powers over these changes, MAFF can give capital grants of up to 50% for providing, replacing or improving field boundaries. Even so, farmers on such marginal land tend to opt for the cheapest option, namely post and wire fencing and, in areas like the Lake District where stone walls are such a feature, the impact on the landscape in the long-term will be considerable.

Much of the attraction of Snowdonia also stems from the juxtaposition of enclosed farmland, forestry and mountains and here too the landscape of the hill country surrounding the mountains is undergoing considerable change. New reclamation techniques and especially aerial spraying with the chemical Asulam, are enabling previously unused steep and rocky land to be cleared of bracken and farmed. In Meirionnydd, 800 of the 7200 ha of bracken-infested hill land had been sprayed by 1974 (Snowdonia National Park 1975). The pattern of forestry is also altering and, again by 1974, 21 103 ha had been planted by the Forestry Commission in the National Park, 95% being conifers. Most of the planting was done before 1960, but throughout the 1970s the Commission have been acquiring and planting about 100 ha each year. In contrast, most of the 1857 ha of amenity woodland is broadleaved and unmanaged. The great majority is derelict and unfenced and, as in the Lake District, now used mainly for grazing and stock shelter, thus precluding regeneration. The situation has forced the Snowdonia National Park authority to turn to outright purchase, rather than to management agreements, as a solution.

The two National Park studies both indicate that the key land-use problems in hill country are afforestation and the twin issues of farmland reversion and moorland reclamation, conclusions borne out by the authors' own 1976 study of hill country in the Dartmoor National Park.

The Dartmoor study area can be divided into a series of ridges and valleys, running roughly north to south with the highest ridges in the west where they merge into the main part of the moor. The landscape of the valleys and the lower ridges is dominated by farmland, that of the valley sides by woodland, and the higher ridges and plateaux by moorland. This parallel zonation and mixture of land use is illustrated in more detail in Figure 4.9, which also emphasises the dominance of two land uses, farmland and heathland. Table 4.7 shows that farmland occupies nearly 50% of the land and underlines the important point that British National Parks are not made up exclusively of wild landscapes. Indeed it is this transition zone, from lowland to moorland, that attracts over half of the visitors to Dartmoor. Only in the centre and western edges of the area does moorland account for the majority of land use. Woodland, which covers some 17% of the area, is the other major land use and the wooded valleys dominate certain parts of the landscape. There is, however, a general scatter of woodland throughout the area and, of 240 kilometre squares, coniferous

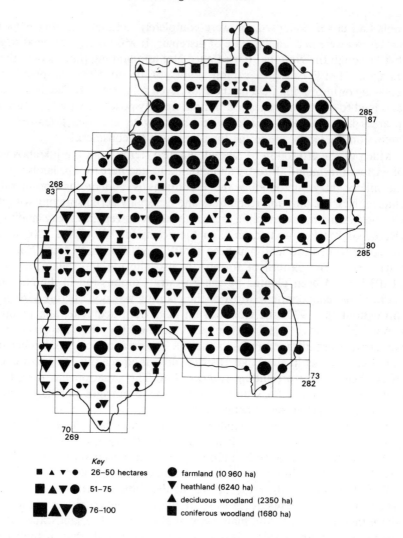

Figure 4.9 Major land uses in the south-east Dartmoor study area 1976. (Authors' survey.)

Table 4.7 Land use in the south-east Dartmoor study area. (Authors' survey.)

Land use	ha	%	Land use	ha	%
farmland	10 960	47·0	developed	1 400	6·0
heathland	6 240	27·0	unused	200	1·0
deciduous woodland	2 350	10·0	water	120	0·5
coniferous woodland	1 680	7·0	orchards	110	0·5

total area 23 060 ha

woodland is absent from only 60 squares and deciduous woodland from only 40 squares. Although developed land covers less than 6% of this part of Dartmoor, this is still equivalent to over half the national average for this type of land and further emphasises the point that British National Parks

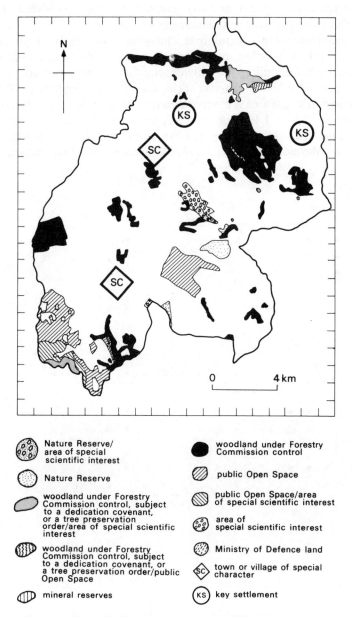

Figure 4.10 Planning constraints in the south-east Dartmoor study area, recorded on the Devon Development Plan, second review (*op. cit.*, Fig. 4.1.)

are places where people live and work, as well as areas of beautiful landscape. Indeed, the charm of this part of the National Park lies in the mixture of farmland, moorland and woodland surrounding the small towns and villages. It is the sort of landscape that has disappeared from many areas of southern England, and any change in its character is thus of great interest to those involved in landscape planning.

The main planning control over changing land use is of course the National Park designation that covers all of the study area, and gives the National Park Committee not only development control powers, but also positive powers of land management (see Ch. 7). In addition to the National Park designation the area is covered by a number of other planning policies, as shown in Figure 4.10. For example, the area contains two villages of special character, North Bovey and Widecombe; and two key settlements where new development is encouraged, namely Moretonhampstead and

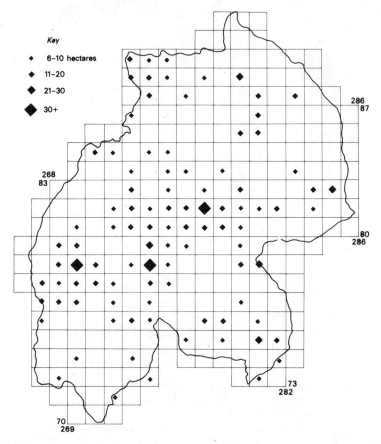

Figure 4.11 Land use changes in the south-east Dartmoor study area. Most change has occurred around the two large moorland areas, and along the major river valleys. (Authors' survey.)

Christow. Most of the woodland is covered by either dedication agreements or Tree Preservation Orders, or is owned by the Forestry Commission. There are three nature reserves and one of these, Yarner Wood, is an important National Nature Reserve. In addition there are three major Sites of Special Scientific Interest. Finally, most of the open moorland is planned as public open space and there is, therefore, a presumption against reclamation. Overall, the area provides an ideal mixture of landscapes and planning policies in which the impact of planning powers can be assessed.

Figure 4.11 shows that there has been a good deal of change on the moorland fringe, in the central and western parts of the area, and that it does not follow very closely the pattern of planning policies. Figure 4.11 only shows the larger changes above 5 ha and it must be remembered that over 90% of all squares show some land use change. Above this groundswell of change, 50 squares with changes of 10 ha or more account for the three main concentrations of change shown in Figure 4.11. In over 150 squares, the majority, only one category of land use change is recorded, and only 14 squares contain more than four, indicating that the pattern is characterised spatially by numerous unrelated and minor changes, but statistically by a few large changes.

This is borne out by Table 4.8 which is dominated by changes between only three categories, farmland, woodland and heathland, together accounting for over 1000 of the total of 1437 ha. Even so the pattern is still a complex one, especially when the two way nature of the changes is taken into consideration. There are seven major flows, as shown:

	Loss	*Gain*	*Net loss/gain*
heathland	881	198	− 683
farmland	351	505	+ 154
deciduous woodland	126	299	+ 173
coniferous woodland	18	287	+ 269

On balance the changes are creating a more diverse landscape, for although the amount of moorland area is being reduced, the landscape value of the area has almost certainly been enhanced by the increases in woodland. It is perhaps worth noting that the rate of change at 6·23% is remarkably similar to that recorded in the east Devon survey, but whether this represents some underlying trend, or is mere accident, can only be judged by relating the changes more closely to planning policies as has been done below:

Direct (compulsory) negative powers over change

farmland to developed land	52 ha
total	52 ha

Indirect (voluntary) positive powers over change

heathland to farmland	473 ha
heathland to deciduous woodland	225 ha
heathland to coniferous woodland	161 ha

Table 4.8 Land use change in the south-east Dartmoor study area c. 1950–76 (ha). (Authors' survey.)

From/to	Orchard	Farmland	Deciduous woodland	Coniferous woodland	Developed land	Heathland	Water	Unused land	Total loss	% of area	Net loss/gain
orchard		18			7			3	28	0·12	− 17
farmland	5		44	32	52	187		31	351	1·52	+154
deciduous woodland		11		88	4	7		16	126	0·54	+173
coniferous woodland	1		10		2	4	1		18	0·07	+269
developed land	1	1						3	5	0·02	+ 66
heathland	4	473	225	161	6		1	11	881	3·82	−683
water										0·00	+ 2
unused land		2	20	6					28	0·12	+ 36
total gain	11	505	299	287	71	198	2	64	1437	6·23	
% of area	0·04	2·18	1·29	1·24	0·30	0·85	0·01	0·27	6·23		

total area 23 060 ha

Indirect (voluntary) positive powers over change—cont.

deciduous woodland to coniferous woodland	88 ha
farmland to deciduous woodland	44 ha
farmland to coniferous woodland	32 ha
total	1023 ha

No real powers over change

farmland to heathland	187 ha
farmland to unused land	31 ha
total	218 ha

Although least change occurred in the 'direct' category, the increase in developed land represents a rate of increase of over 5% on the previous total and, although this is less than the 8·68% recorded in east Devon, it is still a large increase in an area of protected landscape. Furthermore, the new development has not been restricted to the two key settlements, indeed nearly all the increases involve less than 2 ha and are spread out widely over the study area. Nonetheless, as Chapter 6 shows, the rate of increase would almost certainly have been even greater without planning controls.

The vast majority of changes have occurred in those categories where planning policies are indirect. The most significant has been the loss of heathland to farmland, although this has been by no means a one-way flow, with the 187 ha gained from farmland partially compensating for the 473 ha lost. Most of the large losses have occurred on the moorland boundary and have clearly been encouraged by agricultural improvement grants from MAFF. Even so they do not extend the moorland boundary as far upwards as did the nineteenth-century 'newtakes'. Heathland has also been lost to farmland along wet valley bottoms and here again MAFF drainage grants and liming subsidies have clearly played a part. It would appear that the main impact of planning policies has been to encourage the recolonisation of land already cultivated in earlier times and which was already partially enclosed. On Dartmoor, therefore, the loss of heathland to farmland has not conflicted with landscape conservation as seriously as it has on Exmoor and presents little threat to the moorland core. Nonetheless, conservationist planning policies have done little to halt agricultural improvement, and in the whole of the area the wild landscapes, that are the ideal of many proponents of National Parks, are gradually disappearing.

All the other changes in this category involve gains of woodland, mainly from heathland. These changes are now covered by voluntary agreements between the National Park Committee and timber growing organisations, which show where forestry will or will not be welcomed. It should be emphasised, however, that these agreements are only voluntary. Indeed the largest of these changes, heathland to deciduous woodland, is largely accidental and has occurred where heathland has been allowed to revert to woodland at the points where steep-sided river valleys merge into the surrounding plateau upland. This is most welcome from a National Park point of view, but it has usually happened without reference to planning policies. The next largest category, heathland to coniferous woodland, is concentrated in the north-east of the area, where small patches of moorland have been reduced in size as existing coniferous plantations have been extended, and has been more rigorously planned by both landowners and planners. It is the continuation of a past trend and since it does not affect the heart of the moor, there are those who claim that by adding to the diversity of the land use surrounding the moor, it actually enhances its beauty. Furthermore, research by the Forestry Commission has shown that coniferous forests support more wildlife than the upland heaths of Dartmoor (Forestry Commission 1972). On balance, therefore, the change of land use from heathland to coniferous woodland can be said to represent a planning gain in this area of transition from lowland agriculture to moorland.

Coniferous woodland has also gained 88 ha from deciduous woodland, but half of this has occurred in two squares only and represents extensions of existing woodland and is clearly part of a strategic planting plan using Forestry Commission aid and powers. The rest is small-scale and scattered, probably done without grant aid. Nonetheless, it is reasonably certain that the change has been hastened by positive planning powers, and that a potential threat of further loss still lies over the rest of Dartmoor's deciduous woodlands, even though this research has shown the threat at the moment to be more imagined than real. One of the explanations is that deciduous woodland losses have been offset by gains elsewhere. Not only has there been a gain from heathland, but also from farmland (44 ha). Most of this has been in small plots, only one transfer exceeding 2 ha. Clearly, the great majority of the change has been accidental and represents the reversion of unproductive farmland to scrub woodland, indicating the degree to which the management proposals of *New agricultural landscapes* are already occurring spontaneously in the more marginal areas. The last change in this category also represents a gain for woodland, 32 ha of coniferous woodland from farmland. Most of these are extensions to existing woodland and are most likely the result of planned planting using grant aid and Forestry Commission advice.

The final group of land use changes, where planned policies have had little or no impact, includes two types of transfer but is dominated by the 187 ha loss of farmland to heathland. This is widespread and nearly 80 squares record a change of between 1 and 5 ha. It has primarily occurred along wet valley bottoms or on steep rocky hillsides and is clearly associated with marginal land where even the benefit of MAFF grant aid has not made it possible to eke out an economic return. Land that is allowed to return to wilderness may at first sight be thought to represent a planning gain in National Park terms, but such poor land could be put to better use if it were positively managed as wildscape. The loss of farmland to unused land is closely related to the preceding category and almost certainly has occurred for the same reasons. In combination the two changes are closely related to the distribution of transfers in the opposite direction, heathland to farmland, emphasising the marginality of any land use in this transitional area and, maybe, the futility of attempting to draw a hard and fast line around the moorland edge, instead of accepting the inevitability of a continuous process of change. Nonetheless, for practical purposes, it may be sensible for agricultural and conservation interests to agree on an upper limit, beyond which reclamation will not be allowed.

Overall, the Dartmoor study area exhibits a pattern of widespread small-scale and unrelated changes, with a few peaks of much larger change. The most important of these are transfers from heathland to farmland and gains of both deciduous and coniferous woodland from either farmland or heathland. Altogether only about two-thirds of the total is covered by specific planning policies, and at least a quarter is taking place without any planning input. In the sense that the majority of planning policies are based on advice and grant aid as outlined in Chapter 2, this is in line with government policy, but these policies cover the whole spectrum of rural land uses, from farming and forestry to development land, and even in the National Parks they are not fully co-ordinated, although the 1977 National Park plans have made considerable progress in this direction. Nevertheless, it cannot be said that land use change on Dartmoor is being planned as a whole. In spite of all the changes that have taken place in hill country, the intimate landscape that mostly remains is reminiscent of what existed only 10–20 years ago over much of lowland England, making its preservation now all the more valuable, not only for its own sake as a wildlife and scenic resource, but as a buffer zone around the upland areas. It is indeed these areas of transition, between the farmed landscape of lowland England and the remote uplands, that provide prime areas for recreation and, in common with the urban fringe, deserve the most attention from planners in the future. By definition however, the fringe country is always changing and, as the next section shows, it has recently been extended deeper into the uplands.

The uplands

The mixture of land uses found in hill country gradually gives way to moorland with increasing altitude but, in recent years, the boundary has receded upwards as modern methods of farming and forestry have allowed higher tracts of moorland to be reclaimed for more productive uses. The major changes are from moorland to farming and forestry and the reclamation has led to increasingly bitter and heated disputes between the various interest groups concerned. On the one hand, there are MAFF, the Forestry Commission and the farming community, who all want to see the productivity of these barren areas increased. Their views are broadly in line with a statement made by a farmer who resigned from the North York

Moors National Park Committee after it had attempted to prevent reclamation taking place, 'I believe in the farmer's traditional right to grow what he wishes on his land and not to be interfered with. I cannot accept that natural, gradual changes that occur over the centuries have a detrimental effect on the landscape' (Anon 1978). On the other hand, conservationists, the Countryside Commission and the Department of the Environment (sometimes!) regret these changes and either try to prevent them or at least ameliorate their worst effects. Apart from the farmers' traditional right to do what he pleases with his land, at the heart of the debate there lie two main issues: first, how much land has been reclaimed and is still reclaimable and, secondly, what have been the effects of reclamation on the area's ecology, landscape and recreational use. A number of studies have investigated both issues.

In southern Scotland, Parry used Ordnance Survey maps, aerial photographs and field survey to compare post- and pre-1860 patterns of upland land use change (Parry 1976). He demonstrated that 21% of the present moorland area was formerly farmed, and that around 12% was abandoned for cultivation in the 1930s. Further work in the Peak District has improved the survey method and emphasised that the moorland fringe has always been in a state of transition between forest and moor and farmland and that amounts of moorland today are still high compared with many earlier periods (Parry 1977).

More recent changes have been analysed by Davies using parish data from the Ministry of Agriculture Census (Davies 1976). His work compared land use change in the Dartmoor and Exmoor National Parks between 1952 and 1972 and found that grassland had gained ground from both moorland (rough grazing) and crops. The study revealed the extent of moorland loss, particularly on Exmoor. Indeed, it is on Exmoor that most work has been done on land use changes in the uplands and the rest of this section concentrates on that particular example.

Exmoor has had a long history of land use change between moorland, forestry and farmland. The Knight family reclaimed and developed large parts of the southern part in the nineteenth century by ploughing, by introducing new crops and livestock, by building walls, roads and shelters and by planting shelter belts (Orwin 1970). In this century, however, reclamation did not really gather momentum again until after the Second World War, and the interwar period actually saw a great deal of reversion to moorland. It is perhaps ironic that this relatively recent reversion is now exaggerating the amount of reclamation taking place at the present time.

The so-called 'ploughing problem' did not in fact come to prominence until 1962, when a considerable area of open moorland was fenced prior to ploughing and reseeding with grass (Exmoor National Park 1977). In spite of attempts by the National Park Committee to reach an agreement with the farmers to prevent reclamation, fruitless requests to the Ministry of Agriculture to withhold the economically vital grant aid and equally futile requests to the Ministry of Housing to impose Article IV Direction Orders to bring the fencing operation under planning control, the ploughing has proceeded and the land has been reclaimed. The ploughing issue is the reason for Section 14 of the Countryside Act 1968, under which the Minister may make an order requiring a farmer to give 6 months' notice to the local planning authority of any intention to plough up moorland. In addition to this safeguard, applications for grant aid have to be sent by the Ministry of Agriculture to the National Park Committee. It might be thought that these two safeguards would be adequate to provide data about the rate of change and to control it, but unfortunately not all moorland areas are covered by Section 14 orders, only some farmers apply for grant aid, and the reverse process, farmland reversion to moorland, is not included at all. As a result, the current debate about the extent of conversion has lacked precise factual information and the dispute has required special land use surveys.

The first was conducted by the National Park Committee soon after its formation in 1955 and sought to identify what it broadly defined as 'open country'. It was followed by a survey of the principal moorland areas in 1962, which was updated in 1967 and 1974 to identify the most valuable moorland termed Critical Amenity Land. To be included, land has to satisfy at least two of the following criteria:

(a) Basic moorland vegetation, heather, grass, rough grazing,
(b) Remoteness, wildness or openness,
(c) Visual effect,
(d) Existing value or future potential for physical recreation,
(e) Scientific interest,
(f) Contribution to coastal preservation policy.

The datum line for the survey was the newly completed Second Land Utilisation Survey, which showed that 32% of the total Park area was moorland, although not all this land satisfied the above criteria to be included as Critical Amenity Land. Since 1969, farmers have voluntarily notified the Park Committee of reclamation proposals in these areas, but as several important areas of moorland are not defined as Critical Amenity Land, the data still do not provide a complete record of moorland change.

As a result, further surveys have had to be conducted. Sinclair compared the results of the Second Land Utilisation Survey of the mid-1960s with the National Park's survey of open land in 1955, and found the rate of moorland loss to be 1·7% per annum, which extrapolated would lead to the complete disappearance of all moorland by 2001 (Exmoor Society 1966). However, the value of such theoretical exercises is limited, for not only has most of the easily reclaimable land now been ploughed up, much of what remains is either in public or multiple ownership, or is common land, and far less likely to be threatened.

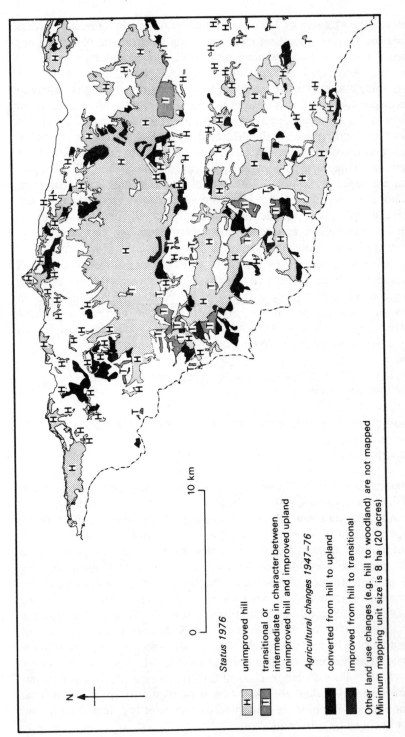

Figure 4.12 Unimproved hill land in central and western Exmoor 1976, and change between 1947 and 1976. Large areas at the margin of the hill land area have been reclaimed. (Porchester, Lord 1977. *A study of Exmoor.* London: HMSO.)

Status 1976

H unimproved hill

I transitional or intermediate in character between unimproved hill and improved upland

Agricultural changes 1947–76

■ converted from hill to upland

▬ improved from hill to transitional

Other land use changes (e.g. hill to woodland) are not mapped
Minimum mapping unit size is 8 ha (20 acres)

0   10 km

N

Since the debate is mainly concerned with future reclamation, two further studies have been conducted. Davies, in a 1976 survey of 41 farms in the Critical Amenity Area, found that the rate of reclamation was accelerating, from a total of 60 ha before 1950, to 409 ha between 1950 and 1965, and 486 ha between 1966 and 1976. A further 1271 hectares, over 40% of remaining rough grazing land, were considered to be reclaimable on physical grounds alone, but when the levels of support grants and subsidies were taken into account then lower levels were forecast (Davies 1977). It is important to note that the definition of what is potentially reclaimable land depends as much on government support for hill farming as it does on the physical characteristics of the land and the fluctuating market prices for hill livestock.

This point was confirmed by Lord Porchester's *Study of Exmoor*, which was conducted in 1977, after the ploughing up controversy had become so acrimonious and divisive that it had started to undermine the work of the National Park Committee (Porchester 1977). Porchester's study concentrated initially on investigating the exact pattern of past changes, and used a combination of 1947 air photographs, Sinclair's 1964 Survey carried out as part of the Second Land Utilisation Survey, and specially flown 1977 air photographs to present the first complete and accurate record of 30 years of land use change as shown in Figure 4.12. The study found that the 24 000 ha of moorland that existed in 1947, had been reduced at an annual rate of 160 ha per annum, and that 4000 ha had been converted to agriculture and 1000 ha to forestry. The Ministry of Agriculture assessed the potential for future change from a detailed field survey, maps and air photographs, using vegetation, gradient, irregularity of the surface, and wetness, as the key variables, and concluded that 64% of the remaining moorland in the Critical Amenity Areas was potentially reclaimable. Lord Porchester's recommendations concerning the management of this land and government reaction to these recommendations are considered further in Chapter 7.

Reclamation of moorland is not, however, confined to Exmoor, it is a live issue in other upland areas as well, notably the North York Moors and Peak District National Parks. Furthermore, in spite of the presumption against moorland reclamation, afforestation is also an important issue in many upland areas. Although most National Parks now have voluntary agreements with the forestry industry, outlining areas as either suitable or unsuitable for planting, these agreements only date from the mid-1960s and the results of previous plantings will still be in evidence for at least the rest of the century. Moreover, many of the upland areas most suitable for forestry lie outside the National Parks, especially in Scotland.

In common with reclamation, afforestation often represents a reversion to a previous land use pattern and, in many areas, it more closely corresponds to the climax vegetation that would exist without human interference, except that instead of the native Scots pine or birch, spruces

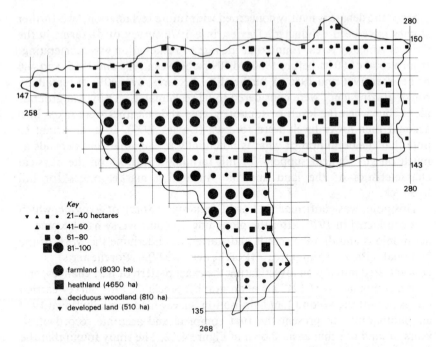

Figure 4.13 Major land uses in the western Exmoor study area 1976. (Authors' survey.)

are now the dominant species. Nevertheless, as Figure 4.7 reveals, the Scots pine still accounts for 14% of woodland and is still Britain's second most common tree. However, because afforestation is usually undertaken over a larger area and on a much longer time scale than reclamation, more care is normally taken to reconcile the new land use with recreational and conservation objectives. In the authors' study of most of the Devon section of the Exmoor National Park, for example, afforestation was seen to be far less of a landscape problem than reclamation. This survey was undertaken in 1976, and of the four areas described in this chapter it is the most remote and the one least affected by pressures for residential, economic or even tourist development. Nevertheless, it has a history of continuous land use change, notably in the transition of land use between moorland and farmland, and the total moorland area has fluctuated wildly over the years. Automatic continuation of the present landscape cannot be taken for granted, as the work by Sinclair, Davies and Porchester, described earlier, showed.

The study area is bounded to the north by the Bristol Channel and the line of high steep cliffs that give Exmoor much of its character. The cliff face is breached by three major valleys: at Combe Martin in the extreme west, at

Table 4.9 Land use in the western Exmoor study area. (Authors' survey.)

Land use	ha	%	Land use	ha	%
farmland	8030	55·5	unused land	370	2·5
heathland	4650	32·1	coniferous woodland	70	0·5
deciduous woodland	810	5·6	water	30	0·2
developed land	510	3·5			
			total area 14 470 ha		

Hunters Inn and at Lynton. These valleys are very deep and steeply sided and are mostly wooded and curve southwards into the plateau heartland of the central moor. Between these narrow winding valleys, the land surface rises south-eastwards as a gently sloping plateau, which can be divided into three roughly parallel land use zones (Fig. 4.13). A narrow band of coastal moorland rarely more than 1·6 km wide; a broad band of mixed farming over 8 km wide in places, and stretching in a narrowing half moon from Parracombe in the west to Brendon in the east; and the moorland proper.

The reality of the cynical comment about Exmoor, that it is an ex-moor and not the moor of the River Exe, is demonstrated by Table 4.9 which shows that over half of the study area is given over to farmland and less than one-third to moorland. As the two uses combined cover 88% of the area, it is not surprising that the transition between heathland and farmland has been, and continues to be, the major theme in the pattern of land use change. A surprising feature of Table 4.9 is the very small area of coniferous woodland, and the only average amount of deciduous woodland. As might be expected, the percentage of developed land is the lowest of all the four areas. The overall mixture of land uses combines to produce a very attractive landscape, probably because the sloping relief allows moorland, farmland, woodland and the coast to be seen together from many viewpoints and research has suggested that diversity of view is the most important factor in explaining landscape beauty. Although, as Figure 4.13 shows, the mixture of land use types is less marked than in either east Devon or Dartmoor, it is hard to reconcile the large and expanding amount of farmland with Dower's definition of a National Park as an area of 'relatively wild country' (Cmd 6628 1945).

Furthermore, Figure 4.14 reveals only one planning policy aimed at the retention of the 'wild landscape', the land intended for 'public open space' with the stress on the word *intended* rather than *designated or purchased*. Also, the large area included in the 'Coastal Preservation Policy' is relevant to development control powers, rather than to the regulation of more general land use change, even though the powers have been widened as a result of the 'Heritage Coast policy' that postdates the County Development Plan. In addition, much of the coastline is owned by the National Trust and so protected from change. Development pressures are officially catered for by Key Settlement status for Parracombe and Coastal

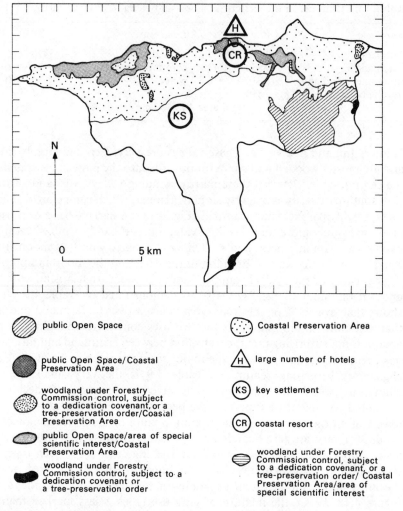

public Open Space

public Open Space/Coastal Preservation Area

woodland under Forestry Commission control, subject to a dedication covenant, or a tree-preservation order/Coastal Preservation Area

public Open Space/area of special scientific interest/Coastal Preservation Area

woodland under Forestry Commission control, subject to a dedication covenant or a tree-preservation order

Coastal Preservation Area

large number of hotels

key settlement

coastal resort

woodland under Forestry Commission control, subject to a dedication covenant, or a tree-preservation order/ Coastal Preservation Area/area of special scientific interest

Figure 4.14 Planning constraints in the western Exmoor study area, recorded on the Devon Development Plan, second review (*op. cit.*, Fig. 4.1).

Resort Status for Lynton. Most of the moorland area is now subject to the voluntary agreement on afforestation, restricting new forestry to non-moorland sites, but a disappointingly small total of existing woodland is under the control of dedication agreements and there must be long-term doubt over the future of the broadleaved woodlands flanking the main river valleys that give this part of Exmoor much of its character.

Figure 4.14 does not show the full range of planning policies covering Exmoor, the one major omission being the boundary of the Critical Amenity Land, where moorland reclamation is resisted by the National Park Committee. There are, however, no absolute powers, other than

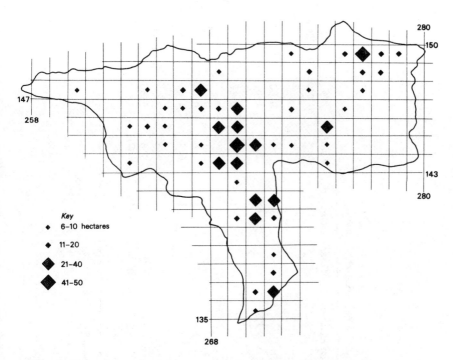

Figure 4.15 Land use changes in the western Exmoor study area. Most change has taken place around the moorland edge, and along the north-eastern coast. (Authors' survey.)

persuasion or expensive compensation, to prevent reclamation and this seriously undermines the policy's effectiveness. Two other omissions from Figure 4.14 are the areas of moorland owned by public agencies, who, as a result of Section 11 of the Countryside Act 1968, can be considered to be less likely to indulge in reclamation, and the areas of common land. Together these cover over 8000 ha in the National Park but, except for some National Trust land, most is in the Somerset section and outside the boundary of this study area.

Figure 4.15 shows that not only is the pattern of land use change more concentrated than in other study areas in Devon, the amounts are also greater. For example, no less than 10 squares exhibit a change of 30 ha or more, indicating that they do not reflect the piecemeal and *ad hoc* changes generally taking place elsewhere, but are instead calculated and premeditated changes on a major scale. The detailed pattern revealed by the field survey maps brings out two important points. First, there are large areas where no change has occurred and is indeed unlikely to occur. For example, the 6% or so of land that has changed in the recent past is the same 6% that has changed many times before and represents fluctuations at the boundary between two types of major land use. Secondly, there are

Table 4.10 Land use changes in the western Exmoor study area c. 1950–76 (ha). (Authors' survey.)

From/to	Orchard	Farmland	Deciduous woodland	Conifer-ous wood-land	Developed	Heathland	Water	Unused	Total loss	% of area	Net loss/gain
orchards		1							1	0·00	0
farmland	1		21	6	23	124		68	243	1·67	+342
deciduous woodland		5		9	1	2		1	18	0·12	+ 71
coniferous woodland			1						1	0·00	+ 35
developed land						1		5	6	0·04	+ 33
heathland		574	61	19	15			5	674	4·65	−547
water									0	0·00	0
unused		5	6	2					13	0·08	+ 66
total gain	1	585	89	36	39	127	0	79	956	6·60	
% of area	0·00	4·04	0·61	0·24	0·26	0·87	0·00	0·54	6·60		

total area 14 470 ha

many theoretically possible transfers which have registered no actual change at all.

Table 4.10 confirms that land use change in this area is both spatially concentrated and limited to a few types of transfer. One transfer alone, heathland to farmland, accounts for no less than 574 ha of the total change of 956 ha. If other losses from heathland are included, namely the 80 ha lost to woodland and the 20 ha to all other uses, the heathland loss total rises to 674 ha. The flow is not, however, totally negative, for heathland has gained 124 ha from farmland and its closely related category, unused land, has gained 68 ha from farmland. Heathland in one way or another is thus involved in no less than 827 of the total 956 ha of change. This all reflects the marginality of farmland and indicates that small changes in economic circumstances can lead to widespread transfers of land use between farmland, heathland and woodland. In the post-war period, the balance of advantage has moved towards farmland, and a net loss of heathland of 547 ha from the former total of 5197 ha, represents a loss of 10%.

The dominance of heathland change is further confirmed by a closer analysis of Table 4.10, which reveals that 49 of the 72 potential transfers between land use categories recorded no change at all, and that of the 23 that did, only 4 showed a change in excess of 23 ha. Further analysis by degree of planning control only serves to confirm these conclusions:

Direct (compulsory) negative powers over change
farmland to developed land	23 ha
heathland to developed land	15 ha
total	38 ha

Indirect (voluntary) positive powers over change
heathland to farmland	574 ha
heathland to deciduous woodland	61 ha
farmland to deciduous woodland	21 ha
heathland to coniferous woodland	19 ha
total	675 ha

No real powers over change
farmland to heathland	124 ha
farmland to unused land	68 ha
total	192 ha

In the case of compulsory powers, the net conversion to developed land, 33 ha, is the second lowest after mid-Devon, in both area and percentage terms, and would have been the lowest, but for large areas of heathland, 15 ha, being converted to car parks or roads. The lack of pressure for development is reflected by the very low rate of residential planning applications in the area, which for the period 1964 to 1973 barely exceeded one application per hectare. This is only partly due to the National Park designation, for the rather bleak hillsides of Exmoor and its relative remoteness make it a less attractive area for settlement than either east Devon or Dartmoor. Nonetheless, planning policies have been able to limit

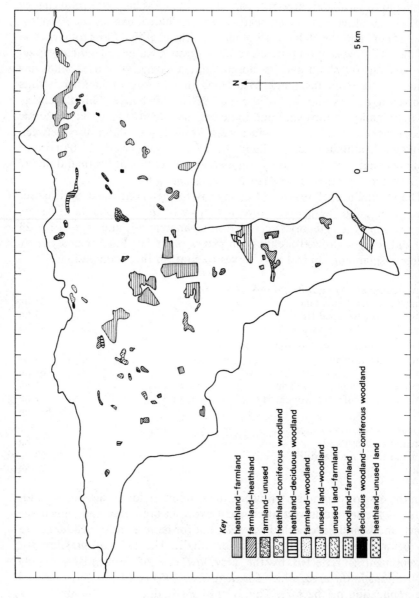

Figure 4.16 Land use changes in the western Exmoor study area. The more detailed pattern of the traditional land use change map can be compared with the more quantitatively based procedure followed in Figure 4.15, which is computer derived (Authors' survey.)

Key
heathland–farmland
farmland–heathland
farmland–unused
heathland–coniferous woodland
heathland–deciduous woodland
farmland–woodland
unused land–woodland
unused land–farmland
woodland–farmland
deciduous woodland–coniferous woodland
heathland–unused land

N

0   5 km

development to existing sites, notably around the fringe of Lynton, and if success is measured by the rather narrow criterion of restricting growth to a few centres, development control has been more successful here than in any of the other three areas.

The majority of change has occurred where indirect planning policies are dominant. In the case of heathland to farmland, by far the biggest single transfer, the grants payable by the Ministry of Agriculture are known to be a considerable incentive, and although Ministry advisers mention the environmental aspects of reclamation as part of their advice to farmers considering improving moorland, it is unlikely that this has halted any but the most marginal reclamation schemes. Another factor since Britain's entry to the EEC has been the grants available under the 'less favoured areas' scheme, which includes moorland. The EEC has, however, agreed to waive these in the case of Exmoor, although in parts of Europe they are used more positively to maintain attractive landscapes, as for example in the German Alps. Figure 4.16 shows the changes in greater detail and, because of their nature, a traditional land use change map is used as well as the grid square approach adopted elsewhere. Figure 4.16 clearly demonstrates that heathland to farmland change occurs in a few concentrated areas, mainly at the moorland edge and along the coast, rather than being scattered throughout the area. Even so, heathland to farmland transfers are found in over 35 squares. The vast majority of change has occurred in a few areas and a detailed inspection of Figure 4.16 and the 1:25 000 source map can lead only to the conclusion that reclamation, with the aid of Ministry advice and grant aid, is an integral part of farm development on Exmoor.

As has been pointed out already, these present changes are the continuation of trends that began in the late 1930s, when the heathland advance was halted by the wartime ploughing up campaign. Marginal areas like this have always been susceptible to land use change, due to outside factors such as agricultural prices, new techniques of husbandry, or population pressure. The last 30 years have seen the emergence of two additional elements: landscape protection in the form of National Parks, and government-induced agricultural expansion. National Park designation has brought a new stability to the fluctuating situation in the form of a general commitment by central government to retain the wild landscape, and a perhaps naïve assumption that the moorland boundary ought to be anchored at the arbitrary and accidental place that it happened to occupy on the day of designation.

Previous land use studies have shown how the moorland edge fluctuates and how difficult it is to freeze it at any one point in time. In the 1930s, admittedly a low point for agriculture and arable cultivation, the core moorland area shown by the First Land Utilisation Survey extended to much lower altitudes and divided the present wide belt of farmland on Exmoor into islands of farmland and surrounding heathland. In 1962, the Second Land Utilisation Survey showed that the wartime ploughing campaign and post-war agricultural grants and subsidies had substantially reduced the moorland area, particularly in the central belt and along the coast. In 1976, as shown in Figure 4.12, the moorland area, so solid in 1932, had become fragmented into four more or less isolated areas, and it is only acquisition by the National Trust that has prevented further reclamation along the coast. It is clear, therefore, that the government policy of agricultural expansion has taken precedence over the somewhat contradictory policy for the retention of open and wild landscape and if further conflict is to be avoided, then there will have to be agreement that the moorland edge is permanently a zone of transition.

The other three changes all involve a gain of woodland. Most of this is scrub invasion of heathland where the wooded river valleys meet the plateau moorland, and it is doubtful if planning policies have had any real impact here at all. Farmland to deciduous woodland similarly reflects scrub invasion of poor quality farmland along valley sides. Only in the case of heathland to coniferous woodland are two large planned plantations found, reflecting the provision of forestry grant aid. In recent years, however, the voluntary afforestation agreement has dampened demand for afforestation in sensitive areas and can be said to have had an important indirect effect. It would also appear that felling licences have acted as an effective deterrent to clearance of the broadleaved valley woodlands, although the very poor quality of much of the timber may be an equally cogent explanation.

In the final category, where no real powers or policies exist, a considerable amount of change has taken place. The change is, however, limited to farmland loss to either heathland or the closely related unused land. Most of the reversion has occurred on small, steep or wet sites and the pattern is in direct contrast to that of reclamation, occurring on a large number of widely scattered, small sites, as opposed to the concentrated large sites where reclamation has been pursued. In other words, reversion is the result of either accidental neglect, as reflected in the numerous instances found along steep valley sides, along wet valley bottoms and in awkward field corners, or overambitious reclamation schemes as witnessed by the large areas of reversion in the moorland core. It is only in such cases that the two processes are related, most reversion is beyond the scope of any planning policies, either direct or indirect.

Overall, it is indirect planning policies that have had the most effect on Exmoor, particularly those aimed at the expansion of agricultural production. Landscape protection policies have been largely ineffective when it came to preventing reclamation, but have been reasonably successful in protecting Exmoor from development. Woodland totals have risen, partly by accident and partly by positive inducement.

Change in a marginal area like Exmoor is inevitable, and planners have now to decide where the limits of marginality lie. It is unfair for conservationists to claim the amount of moorland in 1932 to be a base line, for the area then was atypically large. The real issue at stake is whether present reclamation has gone past the long-term limit of sensible agricultural practice and, if so, how can the moorland edge be anchored in a manner acceptable to both the farming and landscape lobbies. As landscape and agricultural planners have failed to agree on this issue, one is led to conclude that rural planning policies as a whole have been remarkably counter productive, not only on Exmoor but in the uplands as a whole. Apart from very high land and the extensive upland areas of northern Scotland, most of Britain's uplands have traditionally been marginal land. A mix of forestry, farmland and moorland probably represents a more balanced use of the upland resource than moorland alone (Forestry Commission 1971) and recent changes may thus represent a return to a more satisfactory ecological *status quo* and are not as unwelcome as some commentators would suggest. It is a balance of land uses that should be planned for, not a monoculture based on one type of land use.

Summary of the case studies

It is now possible to draw together the findings of all the four Devon studies by the authors and make a preliminary assessment of the effect of planning powers on land use change. In all the areas the amount of change is considerable. Out of a total of 60 920 ha, change has taken place on 3585 ha but, as Table 4.11 shows, the changes are by no means uniform and the variations demonstrate the value of a comprehensive field survey, in contrast to either sampling procedures or the use of agricultural census data from which only net flows can be ascertained. Looking at Table 4.11 in more detail, the first impression is of a remarkable similarity in overall change between the four areas, the percentage only varying from 5·67% to 6·60%. This suggests that for a variety of physical and economic reasons, there is a maximum amount of land use change that can be undertaken in any one period. The table also underlines the two-way nature of many of the changes and how much has taken place on land that is transitional between two uses, reflecting another chapter in a fluctuating situation rather than a totally new situation.

If individual land uses are considered, each shows a consistent pattern of either net loss or net gain. The one exception is farmland, which has three net gains and one net loss, the odd area out being east Devon where there is great pressure for development on the urban fringe of Exeter. Elsewhere, farmland has consistently shown a net gain. Other net gains are shown by woodland, developed land and unused land. The total net gain of developed land, 336 ha, is heavily concentrated in the east Devon study area but represents a 0·55% rise on the total area studied (61 000 ha), a figure well below the national average. Woodland has shown a much larger net gain of 820 ha, three-quarters coming from heathland and the rest from farmland. The 223 ha gained by unused land is less welcome since it is mainly

Table 4.11 Major land use transfers in the four study areas (ha). (Authors' survey.)

From/to	Orchard	Farmland	Woodland	Developed land	Heathland	Unused land	Loss
orchard	— — / — —	114 18 / 102 1	10 0 / 12 0	7 7 / 9 0	0 0 / 0 0	0 3 / 3 0	131 28 / 136 1
farmland	18 5 / 4 1	— — / — —	95 76 / 32 27	184 52 / 23 23	54 187 / 5 124	88 31 / 26 68	439 351 / 90 243
woodland	0 1 / 0 0	12 11 / 24 5	— — / — —	5 6 / 0 1	0 0 / 2 2	12 0 / 3 1	29 18 / 29 9
developed land	0 1 / 0 0	0 1 / 10 0	0 0 / 1 0	— — / — —	0 0 / 0 1	8 3 / 1 5	8 5 / 12 6
heathland	0 4 / 0 0	65 473 / 209 574	126 386 / 24 80	29 6 / 0 15	— — / — —	0 11 / 0 5	220 880 / 223 674
unused land	0 0 / 0 0	0 2 / 2 5	0 26 / 2 8	0 0 / 0 0	0 0 / 0 0	— — / — —	0 28 / 4 13
gain	18 11 / 4 1	191 505 / 347 585	231 488 / 71 115	225 71 / 32 39	54 187 / 7 127	108 48 / 33 79	827 1310 / 494 946
net loss/gain	−113 −17 / −132 0	−248 +154 / +257 +342	+202 +470 / +42 +106	+217 +66 / +20 +33	−166 −693 / −216 −547	+108 +20 / +29 +66	6·24* 6·23* / 5·67* 6·60*

$$\frac{A\ |\ B}{C\ |\ D}$$ A east Devon   B Dartmoor   C mid-Devon   D Exmoor   *Percentage change

Note: Totals do not exactly correspond to other tables due to aggregation and elimination of some land use categories.

accounted for by redundant railway lines, disused quarries and rundown farmland, and, although it is a habitat gain of wildland, it is mostly not being managed so as to realise its full conservation potential. The loss categories are accounted for by two land uses only; orchards and heathland. The orchard loss of 296 ha has been so severe that few are now left. Heathland losses have been even larger in area if not in proportion to their original total; the gross loss of 1997 ha is over five times the gain of 375 ha. The impact of this varies, being most severe in percentage terms in mid-Devon where virtually no heathland is left, and least severe in Dartmoor where substantial areas of heathland have been left unreclaimed. In east Devon and Exmoor the amount of heathland has dropped critically near to the maximum amount required for the areas to qualify as open moorland landscapes.

The spatial variation of the changes both between and within the areas is also interesting. East Devon has experienced the most diverse pattern of change and also the most spatially widespread. Change on Dartmoor is also varied but more concentrated into certain marginal areas where farmland, woodland or heathland meet. In mid-Devon the trend is strongly towards an increased area of farmland, leading to a very simple land use pattern. In Exmoor the conversion of heathland to farmland has dominated all other changes. In all four areas, most large changes have been concentrated along marginal boundaries where one dominant land use gives way to another.

The net effect of all these changes has been to produce a lowland landscape dominated by farmland, and to move the zone of land use transition from farmland to heathland further upwards into the hills. At the same time this transitional zone between farmland and heathland has become more diverse, due to gains of woodland in the form of both planned planting and natural regeneration. It is fortunate that there has been an increase in land use diversity in this transitional zone, for it is the area already heavily used for recreation and one well served by a network of roads and services. However, if the zone moves further upwards it will extend above the existing settled landscape and deprive most car borne visitors of a roadside contact with open moorland and transitional farmland. Recreational pressures are bound to increase as a consequence of the expansion and simplification of lowland farmed landscapes that have rendered much of the lowlands unsuitable for informal recreational use.

From the evidence presented above it is possible to make three general statements about land use change in the countryside:

(a) The area of developed land is continuing to expand, but around existing settlements rather than in the open countryside;
(b) Farmland is being simplified and also extending upwards into previously marginal areas;
(c) The farmland–moorland fringe area is also extending upwards, but is becoming more diversified as a result of the larger areas of woodland.

Table 4.12 Land use change in Devon outside the national parks *c.* 1900–78 (ha). (Devon County Planning Department and Nature Conservancy Council 1979. *The Changing face of Devon*. Exeter: The Department.)

From/to	Orchards	Cultivated	Deciduous woodland	Coniferous woodland	Built-up	Rough grass/heath	Scrub	Loss	Net change	% of area	Authors' survey net change
orchard		2 391	162	13	167	103		2 386	− 2 551	0·45	− 262
cultivated	275		1 040	1 035	166	5 775	317	8 608	+ 23 236	4·12	+ 505
deciduous woodland		1 122		1 830	53	384	75	3 464	+ 2 350	0·42	+ 820
coniferous woodland		338	306		42	370	165	1 221	+ 6 798	1·20	} + 820
built-up									+ 664	0·12	+ 336
rough grass/heath	10	27 946	4 138	5 141	236		1704	39 175	− 32 516	5·76	− 1 622
scrub		47	168			27		242	+ 2 019	0·36	+ 223
gain	285	31 844	5 814	8 019	664	6 659	2261	55 546			

total area 564 500 ha

Table 4.13 Comparison of 'Authors' and 'Changing face of Devon' land use change totals. (Authors' survey and Table 4.12.)

	'Authors'		'Changing face of Devon'	
	ha	% of all change	ha	% of all change
heathland to farmland	1 321	37	27 946	50
heathland to woodland	612	17	9 279	17
farmland to heathland	370	10	5 775	11
farmland to developed land	282	8	166	—
orchard to farmland	235	7	2 391	4
farmland to woodland	230	6	2 075	4
farmland to unused land	213	6	317	1
deciduous woodland to coniferous woodland	119	3*	1 830	3

*Does not include east Devon study area.

An interesting comparison with the figures produced by the authors' study is provided by a joint Devon County Planning Department and Nature Conservancy Council project published in 1979 under the title *The changing face of Devon*. This study compared the land use shown on the early 1900 edition of the Ordnance Survey 1:10 560 map with the land use patterns shown on mid-1960s Ordnance Survey 1:25 000 maps, and was brought up to date in 1978 by student field surveyors employed under the Job Creation Scheme. Although the study is not strictly comparable with the authors' work, the results have been made as compatible as possible for the purposes of this discussion.

Table 4.12 shows the simplified pattern of land use change in lowland Devon between 1900 and 1978 and presents a very similar overall picture to Table 4.11. For example, heathland and orchard show net losses in both studies and all other uses show net gains. Not surprisingly, the amount of change, at 9·84% is half as large again as in the authors' study, because of the longer time span. The table also shows a greater variety of change, with 26 out of a possible 36 transfers recording a change in excess of 100 ha.

When the two surveys are compared by ranking the largest transfers of land use in descending order, as in Table 4.13, a remarkable similarity emerges. There is only one major anomaly, namely, farmland to developed land, but this is entirely due to the *Changing face of Devon* study excluding developed land from its survey. However, two other anomalies are not shown in the table, rough grass or heath to scrub, and deciduous woodland to cultivated land, which recorded totals of 1704 ha and 1122 ha in the *Changing face of Devon* study but only 16 ha and 11 ha in the authors' study. Again an explanation is provided by differences in study method and definition between the two surveys.

Apart from these anomalies, there is a close relationship between the two sets of data, particularly when percentage figures are used. Only heathland to farmland shows a major difference and this is again a reflection of the longer time span involved in the *Changing face of Devon* study and their exclusion of National Park landscapes. It would thus seem that the two studies present a realistic assessment of twentieth-century land use change across a wide range of rural landscapes and there is little reason to suppose that the results do not apply more generally in the United Kingdom.

They can therefore be used as a basis for making some preliminary statements about the effect of planning on the changing rural scene. It would seem that negative controls over new building have not only slowed the rate of rural development, but have also prevented a good deal of sporadic development in the open countryside both by deterring applications from being made and by turning down undesirable applications. Development control can thus be hailed as a moderately successful exercise for even though development has not always occurred in the chosen settlements, it has by and large been restricted to existing developed areas. It would also seem that the majority of land use change has occurred in the arena where planning powers are either voluntary and advisory or depend on financial inducement. This coincides with the British preference for concensus rather than direction, but the success of some policies, notably agricultural expansion, has sometimes seriously undermined the foundations of other policies, notably landscape and wildlife conservation. There is clearly a need for more co-operation between the various authorities in this central area of British rural planning which embraces all the main rural agencies, including MAFF, Forestry Commission, DOE, Countryside Commission, Nature Conservancy Council, National Parks, and local planning authorities. Finally, it appears that a significant amount of change has occurred outside the influence of even informal government powers; the growth of unused and unmanaged land for instance is a luxury that a small and crowded island like Britain cannot afford for long.

These case studies have emphasised that the open countryside can no longer be regarded as 'white land' where change is the province of resource planners only. Significant alterations are taking place in these areas and the evidence presented in this chapter points to the need to establish a wider and more integrated approach to rural planning. So far in the open countryside, it has been the resource planners who have been the most effective in achieving their aims and who have induced the greatest amount of change.

Research is now needed to find out how rural planning policies can best be integrated at the national, regional and local levels. Most important will be the local level, where most land use decisions are taken and the major task here is to identify techniques for assessing the relative value of different rural resources, for example, landscape conservation *vis-à-vis* increased food production. New management techniques are also required to encourage the integration of land uses according to established priorities.

5 Settlement planning

Most direct powers in rural planning stem from the development control fuctions exercised by local authority planning departments and this chapter sets out to define their scope and influence. However, not all controls are vested in the local authorities; central government and its agencies exercise a good deal of indirect influence, particularly in the formation of policy. Each of the major agencies, at both central and local level, will be considered in turn, so that their precise roles may be unequivocally established.

Organisations

Department of the Environment

The government departments mainly responsible for changing patterns of rural settlement are the Department of the Environment (DOE) in England, the Welsh Office in Wales and the Scottish Development Department in Scotland. The DOE was created by amalgamating a number of Ministries in 1970, notably Housing and Local Government and Transport, although the latter became a separate Ministry again in 1976 (Draper 1977). The DOE also acts as an umbrella organisation for a large number of boards, councils and commissions, many of which are concerned with rural planning. The most important of these are the Countryside Commission, the Development Commission, the Council for Small Industries in Rural Areas and the Housing Corporation.

The Department exercises direct influence over the whole land use planning system by instigating new legislation, by ensuring that the existing plannning legislation is properly implemented and by assisting in its interpretation. Informal contacts with local authorities are the chief means by which the DOE is able to promote its point of view, but it sometimes exercises more direct control through the medium of formal public inquiries if a planning issue is of national rather than local concern. The Department is also responsible for settling appeals, where the decision of a local authority is challenged by the applicant.

Despite its co-ordinating role, the DOE does not prepare national plans, although it does attempt to eliminate major discrepancies between local authority Development and Structure Plans and, from time to time, it has also produced long-term national population projections with a view to their being used as a basis for policy formulation (DOE 1971b). The Welsh Office and the Scottish Development Department perform similar functions in their respective countries.

Regional planning authorities

The next tier of control in land use planning used to be the Regional Economic Planning Councils set up in 1964 for the eight English planning regions: the South-West, West Midlands, East Midlands, North-West, North, Yorkshire and Humberside, East Anglia, and the South-East. In 1979, however, all were disbanded although the internal civil service Regional Economic Planning Boards were retained. The Welsh Office and the Scottish Development Department have always performed the role of regional agencies in Wales and Scotland (Eversley 1975). The prime task of the Regional Economic Planning Councils was to produce regional strategies to co-ordinate economic and land use planning mainly with reference to the distribution of population, industry, transport and employment. The strategies also provided a regional framework for local Development Plans and a guide to investment decisions in both the public and private sectors.

Ever since the 1940 Barlow report pointed out the dangers of regional imbalances in population growth and economic opportunity, British governments have attempted to achieve a more equitable distribution of development around the country (Cmd 6153 1940, McCrone 1969). At the national level the Department of Trade and Industry has attempted to steer growth away from the Midlands and South-East to the South-West and North of England and to Wales and Scotland (Central Office 1974). Specific aid for rural areas is provided by the Development Commission and its English agency the Council for Small Industries in Rural Areas, and by the Welsh and Scottish Development Agencies (Northfield 1978). In two of the hardest pressed areas where depopulation and remoteness have bitten deepest, special boards have been created. The Highlands and Islands Development Board, to aid the north of Scotland, was set up in the mid-1960s (Grieve 1973), and the Development Board for Rural Wales set up in the mid-1970s to stimulate industrial growth throughout the Principality (Minay 1977). Both bodies use advance factories and financial aid and expertise to encourage economic growth and, in addition, provide aid to retain and expand services and entertainment. Since the mid-1970s Scotland and Wales also have had additional help from the Scottish and Welsh Development Agencies. The SDA can borrow up to £200 million (Carney & Hudson 1978). The WDA is limited to £100 million (Carney & Hudson 1979a).

There is also the European Community's Regional Development Fund that has been designed to help the economic and social development of

Figure 5.1 Administrative areas of local authorities 1 April 1974. The Counties produce Structure Plans and decide strategic planning applications. The Districts produce Local Plans and decide the vast majority of planning applications. (DOE 1974. *Local government in England and Wales: a guide to the new system.* London: HMSO.)

Europe's peripheral and rural regions. These funds are administered by central government departments like the DOE and Scottish Office and, in 1977, the British share amounted to £180 million, about 28% of the total fund (Glasson 1977, Carney & Hudson 1979b).

Local authorities

Local planning authorities collectively have more impact on the detailed development of the countryside than all the preceding organisations put together. Before 1974, the County Councils had complete control over the planning of towns and rural areas, though not over the County Boroughs that were responsible for most cities with populations of over 100 000. In 1974, as part of a general reform of local government, a two-tier system was introduced in which County Councils exercised control over general policy, while the newly formed District Councils were responsible for its detailed interpretation and execution (DOE 1974b). The new system (Fig. 5.1) made the County Councils much more urban orientated since they now included the former County Boroughs. For example, reorganisation in Devon extended County control to nearly a million people, merely by adding the County Boroughs of Exeter, Torbay and Plymouth. In Scotland the form of the reorganisation was rather different, with half a dozen Regional Councils as a top tier and a second tier of District Councils.

One important exception for planning in rural areas was the National Parks in England and Wales, where the upper tier was made responsible for all planning functions. In all but two cases this meant a committee of the appropriate county councils, with members appointed by the constituent county councils in the National Park and also by the Secretary of State for the Environment. In the Peak District and the Lake District there are separate planning boards, with rather wider powers. As a result there is very much more direct co-ordination between strategic policies and development control in these ten particularly sensitive rural landscapes.

Table 5.1 shows the division of responsibilities between Counties and Districts and underlines that it is the District Councils that now exercise day to day control over population and settlement patterns in the countryside (Central Office 1975). The County Councils produce overall guidance in the form of Structure Plans (replacing the former Development Plans) but they are implemented almost entirely by the District Councils, as the authorities responsible for determining planning applications. Their hand is further strengthened by the fact that they are also the housing authority and, Local Plans, which interpret in detail the general policy aims of the Structure Plans, are their responsibility. This division of responsibility was further confirmed by the Local Government, Planning and Land Act, 1980.

Procedures

Controls over population and settlement change

The most important power available to land use planners is their ability to permit, refuse, or impose conditions on any new development, redevelopment, or change of use that is within the scope of development control (Heap 1976), and the whole question is considered in more detail later in this chapter and in Chapter 6. There are, however, other powers such as the duty to preserve buildings that are listed by the DOE as old or architecturally

Table 5.1 Table of responsibilities of non-metropolitan local authorities (planning and other related matters) from 1 April 1974. (DOE 1974b. *Local government in England and Wales: a guide to the new system.* London: HMSO.)

Function	County	District
Planning		
advertisement control		X
building preservation notices	X	X
conservation areas	X	X
country parks	X	X
derelict land	X	X
development control (a)		X
development plan schemes	X	
listed building control		X
local plans		X
national parks (b)	X	
structure plans	X	
Other related matters		
housing (c)		X
recreation	X	X
transportation planning	X	
tourism, encouragement of	X	X
building regulations		X
litter control	X	X
Footpaths and bridleways		
creation, diversion, extinguishment	X	X
maintenance	X	
protection	X	X
signposting	X	
surveys	X	

(a) Some matters are reserved to the county councils but the district councils receive all planning applications initially.

(b) Two joint planning boards have been set up to administer national park functions in the Lake District National Park and the Peak National Park. For other national parks these functions are administered by a special committee of the county council.

(c) County councils have certain reserve powers.

important. In some cases whole areas may be designated as Conservation Areas, where grant aid and strict planning controls help to preserve the design heritage and the whole fabric of the developed countryside.

Some measure of control over the growth of employment is exercised by the requirement that employers obtain an Industrial Development Certificate (IDC) for all medium- to large-scale factories outside the Assisted Areas of the south-west and north of England, Wales and Scotland. Without an IDC, which is issued by the Department of Trade, planning permission cannot be obtained. There is some evidence that IDC and planning controls have served to stifle organic employment growth in small towns and villages in many parts of southern England, and, partly for this reason the controls were largely abolished in 1980. In other areas, strict interpretation of planning controls, with intensive opposition to even workshop scale industrialisation, has prevented many employers from expanding in rural settings.

A more positive side to planning control is the hierarchy of plans, showing developers where it is likely that planning permission will be given, and where necessary infrastructure for new development, like new roads, schools and water mains will be provided. At the top there used to be plans for the eight English regions and the national plans for Scotland and Wales. They identified areas for growth and extra investment in public facilities and were supposed to co-ordinate the policies of the lower level plans.

In practice it is the County level that is most important and three main types of plan have existed at this scale since 1947: the old style Development Plan, derived from the Planning Act 1947; the new style Structure Plan, derived from the Planning Act 1968; and the Regional Report in Scotland that lies somewhere between a true national plan and the equivalent of an English or Welsh Structure Plan.

Development Plans, which will not be fully replaced in all counties before the 1980s, are basically land use maps showing the type of development that is acceptable in any one place. Although they have had an enormous effect in directing new development to specified locations, they are now considered too unwieldy and dogmatic to accommodate the rapid changes since the 1960s, especially the demands for active public participation (Housing and Local Government, Ministry of 1965). To try to meet such criticisms, the Development Plans have gradually been replaced by Structure Plans, but the changeover has been very slow with only a handful of Structure Plans completed and approved by 1978, although 35, or roughly half the expected total, had been submitted to the DOE.

Structure Plans are basically sets of policy statements, with land use diagrams (not maps) as illustrative, but not statutory background (Housing and Local Government, Ministry of 1970). The whole range of issues covered is presented in Figure 5.2 but local authorities are particularly asked to concentrate on:

(a) The integration of land use and transport planning,
(b) Measures for improving the environment,
(c) The relationship with regional planning,
(d) The resources likely to be available for carrying out the plan,
(e) Social considerations (DOE 1977c).

They are not intended as a detailed guide for development control, as were the Development Plans, but are meant to provide strategic guidance without which control can deteriorate into a plethora of confusing and *ad hoc* decisions, as happened when many Development Plans became hopelessly out of date in the 1960s and 1970s.

Local Plans will normally be prepared by District Councils, and are supposed to provide the detailed basis for the development of those areas

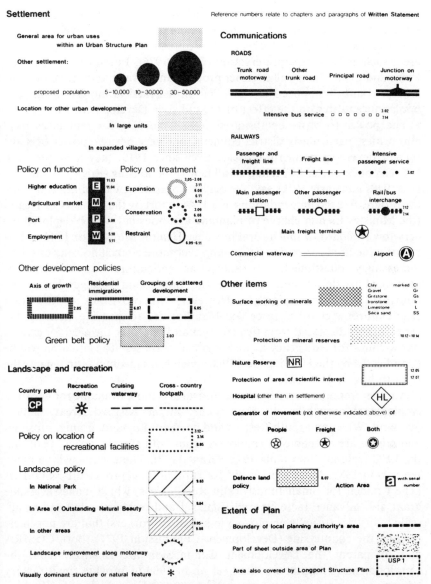

Figure 5.2 Key for the key diagram of a Structure Plan for a rural county. The key diagram brings together the proposals for the development of all major land uses in a county, normally, settlement, landscape and recreation, communications, and other items. (Housing and Local Government, Ministry of 1970. *Development Plans: a manual on form and content*. London: HMSO.)

selected for special attention in the Structure Plan. They will consist of a written statement and a map showing the nature and location of future development and other land use change in the area. Few have been published so far, but it is expected that they will take three main forms: *action area plans* for areas of rapid change, for example an expanding commuter village; *district plans* to prevent, for example the piecemeal development of an historic settlement; and *subject plans* to plan one particular aspect of an area, such as landscape preservation (Fudge 1976). In the future Structure Plans and Local Plans will together produce an overall planning context for a rural county but since they have had little practical impact so far, this book concentrates on the effect of the old style Development Plan.

The power to provide subsidised council housing for rent gives local authorities, particularly district councils, one very positive power over the rate of development in rural areas. Ever since 1919, they have provided between one-quarter and a half of all new rural housing (Rogers 1976). The late 1940s and early 1950s saw the biggest surge of new buildings following the progressive recommendations of a 1944 report by the Ministry of Health (the Ministry then in charge of planning and housing), which deplored the very low standard of much rural housing (Health, Ministry of 1944). Most recently the need to continue providing ever more council housing has been increasingly questioned, particularly as agricultural employment has continued to decline and as more and more slum cottages have been renovated by urban migrants. However, the provisions of the Rent (Agriculture) Act 1976 force local authorities to provide housing for farmworkers displaced from tied cottages and reinforce the need for a stock of dwellings for emergency as well as social purposes. The situation will be complicated by the Housing Act, 1980 which gives medium term tenants the right to buy their houses.

Another important set of positive measures is those available for boosting employment growth in the countryside. In the Assisted Areas, namely Scotland, Wales and south-western and northern England, grants, subsidies and advice are available to employers from both the UK government and the EEC regional fund (Gilg 1976, Grant 1977). These areas will be much reduced in 1982. More specific aid for rural areas is provided by COSIRA (The Council for Small Industries in Rural Areas) which makes available grant aid, advance factories and financial and technical advice to small-scale industries (usually employing less than 20 workers) that are not likely to spoil the countryside (Development Commission 1977). Both COSIRA and its parent the Development Commission are financed from the Development Fund that was first set up in 1909 to provide small factory premises, business management and advice, aid for voluntary bodies who enrich the social and cultural life of the countryside, and general help with new techniques and marketing. Without this aid much increased in the 1970s so as to try and provide some of the 1500 of the 2500 new jobs needed each year in the remote rural areas, the socio-economic structure of the countryside and its physical fabric would have been substantially weakened (Northfield 1977). Indeed the maintenance of thriving communities is now generally accepted as being essential, if a viable rural economy is to be maintained.

As the preceding section has shown, most of both the positive and negative powers over settlement change in the countryside lie with local authorities. These are elected bodies, but direct public participation also still plays a significant part in the planning process (Finer 1974). The government has directed that the public should be involved in the formulation of policies and in the final choice from all the draft alternatives, and that it should participate too in the formal scrutiny of Structure Plans at Examination in Public conducted by the DOE (DOE 1973). Local Plans are to be less formally debated, but are likely to be the subject of more heated discussion, since the issues are more tangible for the individual citizen than the abstract concepts of the Structure Plans. Public participation is very time-consuming, but it does force planners to produce closely argued texts to back up their forecasts. The danger is that it will deflect time and resources away from the more immediate task of day to day decision-making and the problems of implementation. Even when a plan is in operation, it will have failed if both private and public investment are not directed to the desired locations.

Negotiation over the nature of individual developments is usually a process of give and take between the local authority, the developer and pressure groups. The local authority may use its full range of environmental powers to gain tree planting, landscaping and other additions to a residential estate from a developer who initially submitted a proposal for houses only. Conversely, a desperate shortage of new jobs or houses may persuade a local authority to allow developments in areas not outlined for growth by the relevant plan. Plans are not absolute documents, but guidelines to provide a basis for negotiation.

Approaches to policy-making for rural settlement

Development pressures in the rural areas of England and Wales are extremely varied. In some places depopulation and the decline in public services are threatening the very existence of settlements; elsewhere the pressure for development is so great that all vestiges of rural character are in danger of being swamped. Woodruffe has distinguished six different kinds of population change in rural districts in Britain over the 20 year period 1951–71 (Table 5.2) (Woodruffe 1976). They range from those districts with what he calls 'accelerated depopulation', where there has been persistent and growing depopulation over the whole period, through those that grew in population up to 1961 and thereafter declined, to areas of 'accelerated growth', where there has been increasingly rapid growth in population. Thus, although the national trend is one of growth in rural areas, many authorities have to cope with the reverse problem. Frequently, there are significant variations within a single local authority area. For example in the County of Devon there are areas of 'accelerated depopulation' on Exmoor, 'reduced depopulation' on northern Dartmoor

Table 5.2 Trends of population change in rural and county districts 1951–61–71. (Woodruffe, B. 1976. *Rural settlement policies and plans*. Oxford: Oxford University Press.)

	England %	Wales %	Scotland %	Total no.	%
accelerated depopulation	4·4	22·0	34·3	99	14·9
reduced depopulation	7·8	22·0	22·2	89	13·3
reversed depopulation	23·2	23·7	15·7	140	21·0
reversed growth	2·2	6·8	9·1	31	4·6
reduced growth	14·9	6·8	7·6	80	12·0
accelerated growth	47·5	18·7	11·1	228	34·2

and the west coast, 'reversed depopulation' over most of mid- and south Devon, 'reduced growth' in east Devon, and 'accelerated growth' in Torbay and the Exeter–Plymouth corridor (Fig. 5.3).

The patterns are actually even more complicated, for the gross trends do not identify reciprocal transfers of population. In many areas a drift of young people from rural into urban areas is masked by a counterflow of suburban commuters or elderly retired people into the countryside. Such transfers are not necessarily undesirable either socially or economically, but they do place new demands on social and educational services that are difficult to meet in the short-term. Even where the total population is declining, they may result in strong demands for new public investment.

One of the main barriers to an agreed future for rural settlements is the way in which the definition of the problem varies, depending on one's perspective. Nationally, it may appear that there is ample land for accommodating the excess housing and recreational needs of the nation's urban population and that the demands they put on the countryside hold out the possibility of a welcome transfer of resources from the towns to the country. For the rural recipients, however, the picture may be quite different. The popularity of certain locations has meant that thay have become overcrowded and congested, seriously overburdening existing services. The planning authorities in such areas have been under great pressure to formulate policies, designed to keep the urban pressures at bay.

Settlement policies have to deal with a large number of different issues, not all of them amenable to a single, all-inclusive solution and, over the past 30 years, there has been a lack of national objectives and a national strategy for rural settlement. The Countryside Review Committee, in its 1977 topic paper on Rural Communities, concluded that the national approach to the problems of rural settlement had been incoherent and insufficiently sensitive, being in the main a by-product of other policies (Countryside Review Committee 1977b). The major national concerns in the countryside have been the rationalisation and modernisation of agriculture and landscape conservation and although the future of rural settlement has been recognised as integral to both, it has not been treated as a central theme in

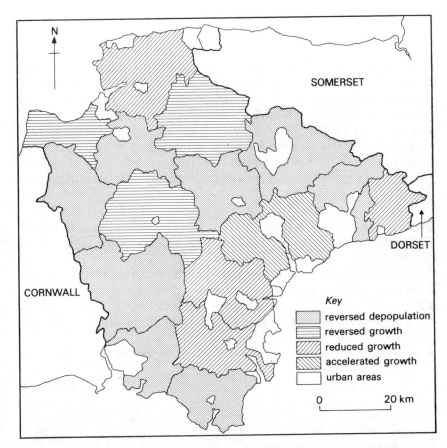

Figure 5.3 Population change in Devon 1950–70. (Woodruffe, B. 1976. *Rural settlement policies and plans in Britain*. Oxford: Oxford University Press.)

its own right. The Countryside Review Committee has no doubt that the future of many aspects of rural life depend on the existence of thriving rural communities. In its opinion economic decline can only lead to a general deterioration in the quality of the landscape and it is therefore vital that new roles be found for villages and small towns.

In the absence of national guidance, local authorities have often looked to the regional plans prepared by the now defunct Regional Economic Planning Councils. For example the South-West Economic Planning Council's 1974 document *A strategic settlement pattern for the South-West* (South-West Economic Planning Council 1974) set out to specify the likely changes in population in the region, in terms of both total numbers and distribution, and thereby provide a framework for the Structure Plan policies that the local planning authorities were beginning to formulate at that time.

The main thesis was that the population of the South-West Economic

Planning Region would grow from 3·8 million to just over 5 million between 1971 and 2001, mainly as a result of in-migration, which it was argued, there would be little chance of reducing through planning policies. Every effort should therefore be made to secure a rate of economic growth in the region sufficient to match the expected population growth, while at the same time doing everything possible to preserve the unique quality of the rural environment. Indeed, the Council argued that the rate of economic growth should be faster than population growth, so as to bring average earnings in the region at least up to the national average. As can be seen from Figure 5.4, the report did not envisage that the population growth would be evenly distributed, although nowhere was it expected to grow by less than the national average. In south-east Dorset, Swindon and Taunton–Bridgwater an increase in excess of 50% was predicted, while Exmoor, the Cotswolds, north Dorset–south-west Wiltshire, and north Cornwall–west Devon were expected to grow by just less than 20% over the 30 year period.

The report was accepted without modification by the Secretary of State as a basis for strategic planning in the South-West, but elsewhere both its methods and conclusions were roundly criticised (Gilg 1975, Cresswell 1975). The unreliability of using trend projections for predicting population

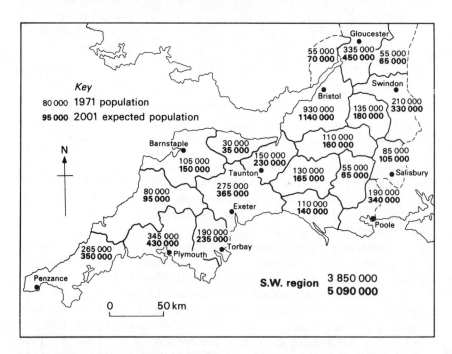

Figure 5.4 Population projections for south-west England 1971–2001. (South-West Economic Planning Council 1974. *A strategic settlement pattern for the South-West*. London: HMSO.)

growth was cited in general and the discrepancies between the forecasts of the Council and those of the Office of Population Censuses and Surveys in particular. The latter predicted a growth rate of between 21 and 25%, as compared with the 32% that the Council was recommending as the basis for strategic planning, a difference of up to half a million people.

One of the report's main theses was that it was not possible to modify significantly migration patterns through development control and other policies, a conclusion welcomed by local authority economic planners, who believed that extra population was the only way to attract new public and private investment. There were many, however, who thought that growth spelt ruination for the rural landscape. The Council for the Protection of Rural England argued that the report ought to have set out goals for improving the quality of life in the region, and also made practical suggestions for resisting any increase in population through migration (Shoard 1974). The cry was widely echoed in many parts of the region itself, particularly Devon, where the prospect of a population increase of as much as 250 000 was viewed with a mixture of horror and near panic in some quarters.

The last four years have seen an apparent vindication for many of the points made by the objectors. Population forecasts for Devon have been sharply downgraded and it is now expected that the population increase in the county will only be between 5 and 10% (46 000–90 000) in the period 1971–91 (Devon County Council 1977). Nevertheless, the importance of the original report for determining general planning strategy should not be underestimated, for although many of its figures have been superseded, its philosophy stands almost intact. The Devon County Structure Plan quotes both the original report and the Minister's supporting letter at length and accepts that the scope for influencing the amount of population growth in the county is only marginal and that the aim ought to be to accommodate growth at the trend rate. The Plan does identify certain critical areas where untoward environmental damage would occur if no policies of restraint could be applied, but argues that for practical reasons the areal scope of such policies should be strictly limited. The implications of this attitude for rural settlement are considerable. Since manipulation at anything other than a local level is ruled out, it follows that towns and villages will be expected to respond to prevailing population trends, rather than dictating their nature and extent.

The *laissez faire* attitude of central government to the future of rural settlements has left a vacuum at the local level. There is no national consensus about the kind of policies planners ought ideally to be aiming for and each local authority has tended to approach the problem in its own way in the light of local pressures, apart from broadly adopting the key village approach and the sieve map technique advocated by the Ministry of Housing in 1967 (Housing and Local Government, Ministry of 1967). The attitude is clearly illustrated by Thorburn in his book *Planning villages* where he states firmly that 'the differences between villages are too great to permit any stereo-typed, rule-of-thumb solution to village planning problems, and one must continually guard against the danger of producing generalisations which are so general as to be useless as guidance in specific cases' (Thorburn 1971, p. 6). He also warns against the adoption of the 'imposed pattern approach'. In his view rigid conceptions of settlement hierarchies and functional thresholds are inappropriate as a basis for planning in the rapidly changing circumstances of contemporary Britain and he accords them a summary dismissal, 'For the time being, therefore, anyone trying this approach is probably doing no more than imposing their own prejudices on a very sensitive environment, with a danger of disastrous results' (Thorburn 1971, p. 45).

Despite such strictures, individual planning authorities have all been faced with one basic problem, the modern rural economy no longer needs the dense network of villages and hamlets which serviced the eighteenth- and nineteenth-century countryside when mobility was measured in a day's return journey of no more than 6 miles. Even though the populations of many country areas are growing, vastly improved personal mobility makes a dispersed network of public and private services both unnecessary and uneconomic. Everywhere the tendency has been towards the centralisation and rationalisation of services in selected locations and restraint on the dispersal of housing, so that it is difficult for the less mobile sections of society to continue to have access to basic services, such as medical facilities and education.

The problem of course is how to decide on an acceptable, or indeed a logical basis for rationalisation and here a number of geographers and economists writing in the 1940s and 1950s have been very influential. Of particular importance is the work of Dickinson on the functional structure of settlements in East Anglia and the extensive studies by Bracey into social and economic provision in the rural settlements of Somerset and Wiltshire (Dickinson 1932, Bracey 1952). Both were essentially empirical in their approach and have been more immediately influential than some of the more theoretical writers on settlement structure. The concept of a hierarchy, with a graduated scale of service provision for each level now underpins virtually all settlement policies and can be traced back directly to their work.

There is considerable variation in the degree to which the policies of individual planning authorities have acknowledged and conformed to such a predetermined pattern and each has naturally tailored its policy to suit the particular conditions of its local area. Nevertheless, in some form or other, the hierarchy is almost universal, as in Devon where settlements were all graded as either regional centres, sub-regional centres, sub-urban towns, key inland towns, coastal resorts, key settlements, or unclassified villages in 1972 (Fig. 5.5).

For rural settlement structure the 'key settlement' is the vital link in the

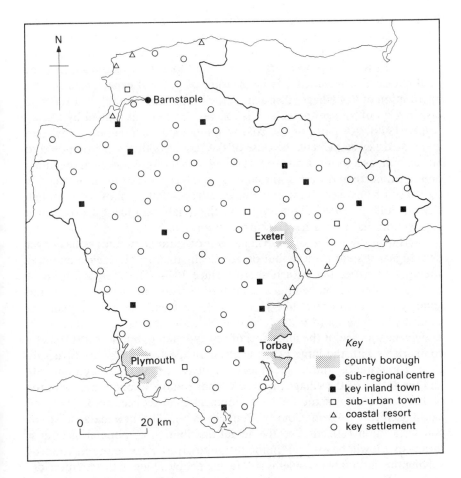

Figure 5.5 Settlement plan for Devon 1972. (Devon County Council 1972. *County Development Plan, second review.* Exeter: The Council.)

chain. It is a concept adopted by a number of authorities in England and Wales, including Cambridgeshire, Durham and Warwickshire, as well as Devon, and depends on the selection of certain villages where investment and growth will be concentrated at the expense of other similar settlements. Cloke has argued that in its purest form key settlement policy is distinguished from other forms of rural planning policy by two features:

'(a) The key settlement itself is planned for comprehensive growth in terms of housing, services and often employment;
 (b) The key settlement policy incorporates an overview of the settlement pattern as a whole and lays special emphasis on the relationship between the key settlement and other settlements served by it' (Cloke 1977b, p. 19).

In Devon the selection of villages for key settlement status was ostensibly based on a complex set of criteria, which included a detailed assessment of both the existing social and economic provision, and the development potential of the village in question. In actual fact it seems certain that the final choice also included a large measure of political expediency, for the distribution of key villages throughout the county does not reflect the actual distribution of villages at all and is too regular to be explained by chance (Cloke 1979). All parts of the county wanted similar numbers of villages designated key settlements because of the resources they were supposed to attract and, as a result, a disproportionate number were chosen in the thinly populated north and west of the county, at the expense of the more densely settled south and east. It raises the interesting general point of whether such elegant policies were in fact any more than elaborate justifications for a contraction that would have occurred in any case?

Interest in the future of the village and of rural communities has never been higher than at present, but paradoxically the rural influence on many county authorities has waned sharply since 1974. The reorganisation of local government removed the administrative distinction between town and country and, in counties like Devon, the spirit and purpose of planning suddenly became more urban orientated. Under the earlier system, rural settlement was one of the main foci of policy-making, but now it suffers by comparison with the large urban centres. In the Devon County Structure Plan, for instance, the needs of Exeter, Plymouth and Torbay dominate and the future of the village inevitably tends to take second place, with the 68 key settlements of the 1972 Development Plan reduced to 38 'Selected Local Centres' (Devon County Council 1979). The new balance is, of course, a fairer reflection of the actual distribution of population, but it means that it will be difficult in the future to divert more resources towards solving the intractable problems still facing people living in the countryside.

Once plans have been formulated, after many years of deliberation, their main task still lies ahead. They must provide a framework to guide and co-ordinate the mass of decisions by local authority development control staff in recommending either the refusal or acceptance of applications to the monthly meetings of the planning committees of local councils. Most research work has so far looked at the formulation of plans and not their implementation. The rest of this chapter and Chapter 6 seeks to redress this balance by examining in some detail the processes of development and its wider effects on rural settlement.

Development control powers

The control of development has been a central feature of the planning system, ever since the then Minister of Housing, Harold Macmillan, changed the main emphasis from positive implementation to negative control in 1952. Development is defined by the planning acts as, 'the carrying out of building, engineering, mining and other operations in, on, over or under land, or the making of any material change in the use of any building or other land'. A broader and more all-embracing definition could hardly be imagined and, were it not for some substantial exemptions, would mean that permission would have to be obtained for any material change of land use, no matter how small or where it was located.

In practice, however, the planning system has become primarily geared towards controlling urban development and rationalising its growth in the countryside. Agriculture and forestry and buildings connected with these uses are all specifically excluded from the definition of development, which means that many of the more important types of rural land use change are outside planning control. The reasons for the predominantly urban emphasis are largely historic, reflecting the pre-Second World War concern with containing the tide of suburban growth and the pressing need after the war to reconstruct the bombed centres of many cities. But it was not just the immediate problems of towns and cities that caused legislators to ignore almost completely the physical planning of the countryside, rural areas beyond the urban fringe were not at that time thought to be under any general threat from development. The use of the countryside for leisure was in its infancy and, in any case, was the preserve of the idealistic and privileged few who could afford the time and had the means to get there. The main problem facing the countryside was not too much development but too little, as Hookway and Hartley have been at pains to point out (Hookway & Hartley 1968). The chief legislative concern was to find ways and means of preventing its premier industry, agriculture, from reverting to the kind of depression that had dogged farming throughout the interwar years. The emphasis was on increased food and forestry production, notwithstanding the mounting pressure from conservation and recreation interests that culminated in the National Parks and Access to the Countryside Act in 1949. The whole idea of either agriculture or forestry having to bear the extra burden of planning control was an anathema in many politically influential circles and this attitude was evident in the early legislation. Only gradually over the past 30 years has the need for strict development control powers in rural areas come to be more widely accepted and there is still much debate about land use priorities in the countryside.

In 1947 the original intention was to plan by consent through agreed plans, which would indicate the kinds of development to be carried out by public authorities, but since 1952 private developers, albeit constrained by development control powers, have come to play an ever more important role. The impact of the legislation has also varied from place to place, in general being much greater in built-up areas than in the countryside. The Town and Country Planning General Development Order (GDO) gives a general permission for a number of types of development, many of which relate to farming and the countryside (DOE 1977b). The effect of the GDO is that developers can assume that permission has been granted for those developments listed in the Order and they therefore neither have to seek permission, or even inform the planning authority of what they propose doing. As far as rural areas are concerned the most important types of permitted development are those contained in Class VI 'Agricultural buildings, works and uses' and Class VII 'Forestry buildings and works', both of which are contained in Schedule 1 of the Order. It is worth quoting the relevant sections so that the full extent of the permissions is absolutely clear:

'Class VI – Agricultural buildings works and uses
(1) The carrying out on agricultural land having an area of more than one one acre and comprised in an agricultural unit of building or engineering operations requisite for the use of that land for the purposes of agriculture, (other than the placing on land of structures not designed for those purposes or the provision and alteration of dwellings), so long as:
  (a) the ground area covered by any building erected pursuant to this permission does not, either by itself or after the addition thereto of the ground area covered by any existing building or buildings (other than a dwelling house) within the same unit erected or in the course of erection within the preceding two years and wholly or partly within 90 metres of the nearest part of the said building, exceed 465 square metres.
  (b) the height of any buildings or work does not exceed 3 metres in the case of a building or works within 3 km of the perimeter of an aerodrome, nor 12 metres in any other case.
  (c) no part of any buildings (other than moveable structures) or works within 25 metres of the metalled portion of a trunk or classified road.
(2) The erection or construction and the maintenance, improvement or other alteration of roadside stands for milk churns, except where they would abut onto any trunk or classified road.
(3) The winning and working, on land held or occupied with land used for the purposes of agriculture, of any minerals reasonably required for the purposes of that use, including:
  (i) the fertilization of the land so used, and
  (ii) the maintenance, improvement or alteration of buildings or works thereon which are occupied or used for the purposes of the aforesaid, so long as no excavation is made within 25 metres of the metalled portion of a trunk or classified road.

Class VII – Forestry buildings and works
The carrying out on land used for the purposes of forestry (including afforestation) of building and other operations (other than the provision or alteration of dwellings) requisite for the carrying on of those purposes, and

the formation, alteration and maintenance of private ways on such land so long as:

(a) the height of any building or works within 3 km of the perimeter of an aerodrome does not exceed 3 metres;

(b) no part of any buildings (other than moveable structures) or works within 25 metres of the metalled portion of a trunk or classified road.'

The effect of this blanket permission is to enable the farming and forestry industries to carry on their operations with little hindrance from planning. Indeed, when the GDO was first made most agricultural buildings and works, with the exception of dwellings, were in practice exempt from planning control. Today, however, the general trend towards larger units means that it is much more common for agricultural buildings to exceed 465 m^2 and therefore require planning consent. It is symptomatic of the slowly changing attitude to development in the countryside that this size threshold has not been raised, although the exemptions remain considerable and, in certain respects, exceed what is technically laid down in the literature. For instance, dwellings although covered by planning controls are often able to get permission by pleading a case for 'agricultural need' and a number of Development Plans actually include an exception clause to this effect in their general policy of prohibiting scattered development in the countryside. Both Samuels and Turrall-Clarke have pointed to the loopholes still available to would-be developers through pleading an agricultural need, despite attempts by the DOE in Circular 24/73 to tighten controls (Samuels 1978, Turrall-Clarke 1976).

One of the more curious features of development control in rural areas is the small number of special controls in National Parks, Areas of Outstanding Natural Beauty and other protected landscapes. Despite the fact that nearly 50% of the land area of England and Wales has now been designated under one or other of these protection orders and the strictures about the need for high standards of development control in the official government guidance to local authorities, there are few extra powers available. The only exception is the Town and Country Planning (Landscape Areas Special Development) Order (Town and Country Planning, Ministry of 1950) that reduces the general exemption from planning controls for agricultural and forestry works and buildings in the General Development Order. In those areas covered, no building other than a moveable structure may be erected, altered or extended until at least 14 days after the local planning authority has been given written notice of the intention to do so. The written statement must include a short description of the proposed building or extension and of the materials to be used, and a plan indicating the site. If within the 14 day period the applicant is notified by the planning authority, he must gain their approval for the design and external appearance before any development is commenced. However, as

Stephenson has pointed out, planning authorities may only act sparingly, 'As regards the exercise of this power, Ministry of Housing Circular 39/67 points out that modern efficient farming practice depends increasingly on buildings, some of which, such as tower silos and grain drying and storage buildings, must of necessity be tall; in the circular the central government asks local planning authorities to bear in mind the needs of the agricultural industry and to use their powers primarily as a means of securing suitable siting and design of farm buildings, rather than refusing permission for their erection' (Stephenson 1975, p. 203). In fact only three Landscape Special Development Orders have ever been made; in the Snowdonia, Lake District and Peak District National Parks. In practice there appears to be some reluctance to use this extra control, the reason probably being the very clear Ministerial directive, that restrictions on agricultural development made under the Order will only be sanctioned in exceptional circumstances. Also, if any local planning authority, whether or not it is part of a National Park, wishes to prevent a particular permitted development from going ahead it can apply to the Secretary of State for an Article IV Direction under the GDO. This revokes most exceptions and a planning application has to be made in the normal way. For an Article IV Direction to be effective however, it is essential for the planning authority to have prior knowledge of the proposed development and, for obvious reasons, such information is hard to come by. Furthermore, the DOE is reluctant to issue orders in all but exceptional circumstances, arguing that they should be an *ad hoc* measure for emergency use.

One final power is the Section 52 agreement procedure that enables a planning authority to enter into an agreement with any land owner for the purposes of restricting or regulating the development or use of land, either permanently, or during such period as may be prescribed by the agreement. There are, however, doubts about the legality of many agreements if they go beyond the strict limits of planning (Tucker 1978). Such management agreements have been made, notably on Exmoor to prevent the conversion of open moorland to other agricultural uses, but their usefulness is somewhat limited. In rural areas the main difficulty arises from the problem of agreeing compensation terms with the owner for loss of development rights. Also, even though it has been possible since 1962 to make Section 52 agreements registrable, thus tying the agreement to the land if it is sold, in practice they have only rarely been concluded because of the damaging effect they could have on the future value of the land (Aves 1976). There are also problems concerning possible charges of corruption and collusion between developers and officers in trading off planning gains for permissions (Jowell 1977, Loughlin 1978).

There are some categories of land exempt from all planning controls, the most notable being Crown and Duchy lands and land owned by government departments. Locally these can be very important, especially in areas of open country. For instance, much of the heart of the Dartmoor National

Park is owned by the Duchy of Cornwall and therefore beyond the control of the National Park Authority. Even though the Duchy usually consults about any proposed development, it is under no obligation to do so. Government departments are subject to a somewhat more formal procedure in that DOE Circular 80/71 requests that they consult, but once again it is up to them and not the planning authority to take the initiative (DOE 1971c).

Development control processes

For those categories of development that actually require planning permission in rural areas, the method of making an application and the process by which it is determined are exactly the same as anywhere else. Anyone may apply for permission to develop land but an application can only be accepted if there is an accompanying certificate, stating either that the applicant owns or is a tenant of the land in question, or that he has given notice at least 21 days previously to everyone who was then an owner or tenant of the land, or that he has been unable to do this but has taken all reasonable steps to try and comply. The certificate must also state that none of the land forms part of an agricultural holding, or that notice of the application has been given to anyone who was a tenant of an agricultural holding forming all or part of the land in question, at least 21 days previously. When considering an application the planning authority must take into account any representations made by either an owner or tenant.

There are two distinct alternatives in the actual determination process. An applicant may either seek an outline permission to decide in principle whether the proposed development may be allowed on the site and then follow this with a detailed application to decide the details of layout and design, or he may apply for both at the same time. Outline permission may be granted with only a description and rough drawings of the proposal, but a detailed permission requires precise plans and drawings. The main reason for the two-stage procedure is that detailed plans are expensive to produce and an applicant could spend much time and money on a proposal, which has little chance of success. An outline application enables the principle of development to be tested cheaply. Increasingly it has also come to be used as a quick means of increasing the value of land, since a site with outline planning permission is obviously a much more desirable commodity than one without. The majority of applications go through this two-stage process but there are a few minor exceptions, the most important of which is temporary planning permission, which covers such things as the temporary use of a building for residential purposes, or the parking of a caravan on a site for a limited period. Such applications are spared the double scrutiny of outline and detailed permission and are determined by a single decision.

Once an application has been made the authority theoretically has 2 months to make a decision and an extended process of consultation begins immediately. The precise details of how the consultations are organised and

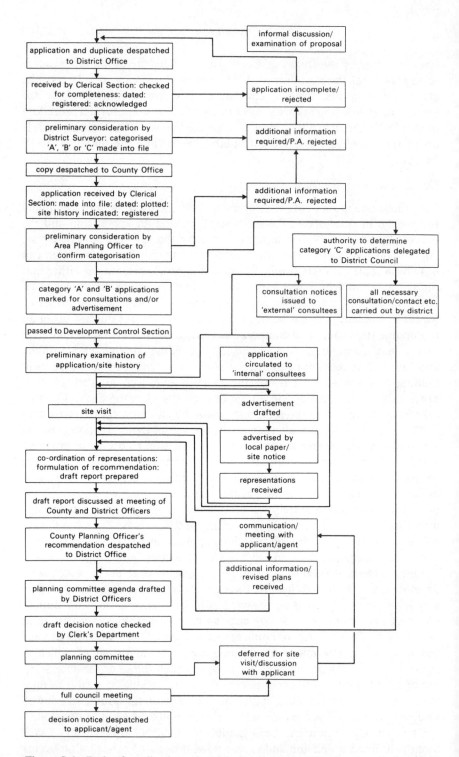

Figure 5.6 Path of applications through the development control system in north Worcestershire. (Joint Unit for Research on the Urban Environment 1977. *Planning and land availability*. Birmingham: JURUE.)

the decision finally reached, vary from one planning authority to another, although the general sequence of events is broadly the same. In a study of planning and land availability in the Midlands the Joint Unit for Research on the Urban Environment has shown the precise details of how applications are determined in north Worcestershire and this serves to illustrate the main stages in the consultation and decision-making process, as shown in Figure 5.6 (Joint Unit for Research 1977).

Ideally, the first contact between the applicant and the planning authority should take place informally before the application is lodged, so that any obvious problems can be eliminated before the formal machinery is set in motion. Large development contractors establish such contacts with the planning authorities in the areas where they intend operating, as a matter of course. Drewett has shown how these firms establish regional sub-offices and one of their major tasks is to strike up a friendly relationship with the planning officers, so as to promote the company's good name (Drewett 1969). Nevertheless, such informal contacts are often not particularly productive because of the fundamental philosophical and ideological differences that exist between planners and developers. The developers' criteria are economic; they know what the market wants and are in business to provide it. The planner on the other hand has wider altruistic considerations and professional responsibilities, many of which are not quantifiable in monetary terms and relate to basic environmental standards; economic considerations frequently assume a subsidiary role. The conflict is inherent in all dealings between developers and planning authorities, but is often particularly acute in rural areas.

Much development is not undertaken by large companies but by small builders and individuals and, although little work has been done on this subject, in these cases there would generally seem to be little contact between the person seeking planning permission and the planning authority before the application is lodged. In 1972 a questionnaire survey was undertaken of 340 private individuals who had made planning applications on their own behalf in east Devon between 1967 and 1972 (Blacksell & Gilg 1974). Less than a third had any direct contact with the planning authority as such, although nearly everyone took some professional advice, either from solicitors, architects or builders. Those that did consult the planning office directly seemed generally to find the help and advice they received less useful than that from other sources. It is also worth noting that this reaction was just as strong from those whose applications were eventually granted, as from those who were refused permission. Effective consultation between planners and developers seems therefore to be strictly limited, but it should be stressed that most of the available evidence dates from the period before local government reorganisation in 1974. Since then planning authorities have made great efforts to bridge the gap more successfully, by making information about proposed developments more freely available and plans more easily comprehensible.

Once a planning application has been correctly made the authority is bound to consider it formally and to come to a decision within 2 months. This has been a basic requirement ever since 1947. Prior to 1974 ultimate responsibility for processing all planning applications lay with the county planning authorities, which meant in most rural areas that planning tended to be a somewhat remote and bureaucratic exercise. Even though most large authorities used to divide their territory into smaller sub-regions for planning purposes and even delegate some planning functions to the urban and rural district councils, planners and planned tended to be somewhat isolated from each other. In Devon, for instance, there were five area sub-committees and for most purposes they were the effective planning authorities in their respective areas; only applications with county-wide significance were referred for discussion to the full planning committee. Indeed the process was even further decentralised, since under delegated powers planning applications were initially lodged with the urban and rural district councils, who considered them and then forwarded recommendations to the area planning sub-committee. Nevertheless, the full county committee remained the ultimate planning authority and inevitably any disputes had to be decided at this level.

Since April 1974, although the county has remained the strategic authority with complete responsibility for structure planning, it now has only very limited jurisdiction over development control. Only if an application is likely to require some direct modification to strategic policy is the county in a position to interfere (Watkins 1979). The new district authorities take most of the decisions affecting individual applicants and are responsible for interpreting broader county policies through the medium of Local Plans and the development control system. The process was further accelerated in 1980, when the Local Government, Planning and Land Act 1980 gave district councils almost complete control over the processing of planning applications, leaving the counties with only applications straddling National Park boundaries and those that involved minerals (Gilg 1980). The only major exception is in the National Parks. Both before and since the 1974 local government reorganisation, the individual National Park Boards or Committees have acted as the planning authority. All applications are submitted to them and they alone have responsibility for determining the outcome. Naturally there is normally extensive provision made for non-statutory consultation as with both the district and county planning authorities, but National Parks do not have to accept them and the details of the arrangements vary from one Park to another.

The more formal division of responsibility between county and district since 1974 has produced a much greater procedural uniformity, but such things as the nature of application forms and internal arrangements are still not spelt out in detail in the legislation. As a result there can be considerable differences between one authority and another. This is important because it can materially affect the nature and extent of consultation, the most time-consuming part of the determination process. Some consultations are statutory, such as referring major departures from the Development or Structure Plan to the Department of the Environment, but others are optional and merely reflect a desire on the part of the planning officer to be as well-informed as possible about the implications of a proposed development. The range of these consultations can be considerable including other departments in the local authority, for example, the engineer's department where roads are likely to be affected, statutory undertakings like the Water Authority, and central govenment departments and allied agencies, such as the Department of the Environment, the Ministry of Agriculture, Fisheries and Food, the Countryside Commission and the Nature Conservancy Council. The importance and nature of these consultations naturally vary, depending on the details of the particular application, but in general, local contacts are more frequent than those with central government or its agencies. Local highway authorities with their responsibility for roads tend to be deeply involved in the planning process, as are the Regional Water Authorities, which have jurisdiction over water supply, sewage disposal and land drainage. The relationships of both these bodies with central government are usually more formal and, unless there is some statutory obligation, it is unusual for there to be any public participation at this stage.

MAFF is an important exception to the general rule. Many proposed developments in the countryside requiring planning permission are in areas where approval would be unlikely in the absence of extenuating circumstances. If, however, an agricultural need, supported by Ministry officials, can be demonstrated then there is a much greater likelihood that the planning authority will make an exception. Indeed, once a genuine agricultural need has been established, a planning authority is almost powerless to oppose the development because it will invariably be overruled on appeal by the Secretary of State. All that it can effectively do is argue about the details of design and siting. This role as arbiter of agricultural need has given MAFF a uniquely powerful position in deciding the nature and scale of development in rural areas. It would be quite wrong to give the impression that the Ministry has only used its influence to by-pass planning policies, using agricultural need as a loophole; it has also intervened to preserve high quality farmland from development. A study made more than 20 years ago in Warwickshire showed with what success they had resisted applications covering more than 5 ha in the countryside (MAFF 1958). Recently it has been further strengthened by the government's reply to the Strutt Report (Advisory Council 1978), which announced that henceforward MAFF would be consulted on all applications involving a loss of more than 2 ha, rather than the previous limit of 4 ha.

Once consultations have been initiated, the next stage is to collect and assess any representations made by persons likely to be affected by the application. If there are objections, the planning officers usually try to mediate before the proposal goes to the committee for a final decision but this is by no means always possible. When a committee is faced with an application to which there is serious opposition, it may try to mediate between the various factions on its own account. Indeed in difficult cases the planning officer may recommend a meeting of his own accord. However, once such consultations are complete the committee has to come to a decision and either grant or refuse the application, although it may choose a tactical delay, hoping that this will force the developer to appeal on those grounds.

Despite all the advice they receive, committees have considerable discretion in their final decision. Harrison has looked in some detail at the various factors influencing development control decisions and comes to some interesting conclusions about the ways members eventually make up their minds (Harrison 1972). He believes that central government is able to exert its greatest influence on local government representatives through legislative action. The effect is to inculcate a set of values, which certainly colour attitudes, but do not necessarily provide much help and guidance in determining difficult applications. The advice given in the 'Development Control Policy Notes' illustrates the point, 'the effect of a proposal on road safety or the beauty of surroundings, the effect on public services, such as drainage and water supply, conflict with public proposals for the same land, the need to ensure that valuable minerals are not sterilized by the erection of buildings over them and many other factors. They must, however, be genuine planning considerations, i.e. they must be related to the purpose of planning legislation, which is to regulate the development and use of land and not some extraneous purpose' (DOE 1974a, p. 4).

Harrison claims that there are four specific ways in which central government has influenced attitudes:

(a) The continuing influence of the containment philosophy, with its emphasis on protecting the countryside;
(b) A consistent belief in physical order as a means of achieving economic and social aims;
(c) The use of development control as a protective measure, independent of social and economic objectives and costs;
(d) A tendency, possibly in order to ensure a consensus about planning, to rely upon established methods and definitions (Harrison 1972).

The first of these is clearly of considerable importance for the present discussion, since it implies a general presumption against development in rural areas. The reliance on precedent has made it difficult to use development control as a mechanism for instigating social and economic changes. Planners and the committees they serve feel deeply bound by previous decisions and these are invariably liberally quoted as precedents for or against coming to a particular decision. As Harrison suggests, this is partly due to a belief in consistency as an important ingredient in producing an image of fair and equitable treatment, but the deeply held belief of most planners that they are concerned with land use priorities and not with other issues such as land ownership and the personal circumstances of applicants is an important subsidiary factor. Whether or not this is a correct or even tenable stance is hard to say, but it does have the effect of removing most planning committee discussions from the immediate political arena. Broad questions of policy rarely arise in development control discussions and, conversely, matters of local detail can sometimes assume undue importance. In the final analysis one cannot but concur with Harrison's conclusion that, 'there is no close relationship between the Ministry and planning authorities' and what contact there is tends to filter one-way only, downwards through the legislation and government circulars.

The main way in which central government becomes directly involved in the development control process at the local level is through the appeals system. An applicant may go to appeal if he is refused planning permission, if he considers the conditions imposed to be too onerous, or if the 2 month time limit for decision has been exceeded. About 10% of refusals go to appeal and the Minister's decision in these cases is binding on local authorities. However, only about 30% are allowed and, if there were persistent disagreement with any one authority, the Minister could apply pressure to enforce greater compliance.

Central government may also become directly involved through the 'calling in' procedure. Any large application, especially if it is a major departure from the agreed Development Plan, may be judged to have implications beyond the jurisdiction of the local planning committee. In such circumstances the Minister has the power, under Section 35 of the Town and Country Planning Act 1971, to call in the application and determine it himself. If he wishes, he may first call a local inquiry under an independent inspector to examine all the issues raised, but he is not bound by the findings and is still personally responsible for the ultimate outcome of the application.

Of the 400 000 or so applications made each year, public inquiries are only held in less than 1% of cases, but their small number belies their growing importance. They are mainly called to examine major applications, such as reservoir and road schemes and, therefore, the decisions resulting from them have a disproportionate impact on the environment. They also exert considerable influence on public attitudes towards planning since it is issues like those covered by inquiries that inevitably tend to come most into the public eye. It is perhaps unfortunate that in the public imagination planning is often equated with major confrontation rather than with a bureaucratic and very routine decision-making process, which is closer to the reality.

Considering the volume of applications and the complex processing procedures, it is surprising that so many are actually determined within the 2 month statutory period. For small-scale applications such as extensions and conversions, which in most areas make up nearly 50% of the total workload, it is not too difficult to comply with the official time-schedule, but applications for new building pose quite another problem. A very detailed study of residential applications in south Birmingham has shown that in each of the 6 years between 1968/69 and 1974/75, the average time taken to finally come to a decision was in excess of 2 months, in some cases considerably so (Fig. 5.7). The authors of this study make three important points about their findings:

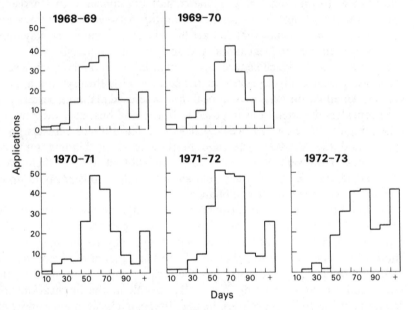

Figure 5.7 Overall time in days from submission of application to determination by Birmingham County Borough. (*op. cit.,* Fig. 5.6).

(a) The distributions are peaked about the 60 day mark, reflecting desire (always less than 50% successful) to process applications within the statutory 2 month period.
(b) The distributions are skewed to the right, with the corollary that measures of central tendency, such as the mean, are a little misleading in that they mask the fact that a small but significant percentage of applications take a very long time indeed to process, sometimes as long or longer than it takes to build a house.
(c) There is no simple or direct relationship between the number of planning applications and the time taken to process them, in part because of the cumulative build up of work load. This is particularly

obvious in 1972–3 when the low throughput in the 60 day period undoubtedly reflects the large carry over from the tail of the 1971–2 distribution (Joint Unit for Research 1974).

It should also be noted that any figures for 1974–5 are unusually distorted, because of the disruption caused by local government reorganisation, a fact that is underlined by the increasing number of decisions being made within the 2 months time limit in the late 1970s. A 10% sample survey of development control performance by the Association of District Councils, among its 333 members, showed that in the year 1977–8, 72% of decisions were made in 2 months, 17% in 3 months and only 10% took 6 months or longer. The main reasons for delay were slow consultations with statutory undertakers and developers, who had often submitted poor applications (Storey 1978).

Table 5.3 The time taken to determine residential planning applications in the period 1964–73 in the four study areas (%). (Authors' survey.)

	East Devon	South-east Dartmoor	Mid-Devon	Western Exmoor
decided in 1st month	14	6	22	15
decided in 2nd month	15	10	8	15
decided in 3rd month	17	13	14	11
more than 3 months	54	71	56	59
total	100	100	100	100

There is some evidence to suggest that delays may still be substantial in rural areas, particularly where protected landscapes are involved. Table 5.3 shows the length of time taken to determine residential planning applications in four areas of rural Devon (Fig. 5.11). Two are in National Parks – Dartmoor and Exmoor – and one – east Devon – is partly within an Area of Outstanding Natural Beauty. Only between a fifth and a third were determined within the statutory period and in all four areas over half took more than 3 months to decide. As in the Birmingham study there is no evidence that volume of work seriously influenced the length of time taken. It seems certain that the remoteness of many of the sites and the sensitivity of the landscapes involved were the crucial factors in delaying the decision.

There is of course nothing magic about the 2 month statutory period; the quality of the final decision, irrespective of how long it takes, is the paramount consideration. Nevertheless, delay is unfortunate for it not only breeds dissatisfaction amongst intending developers, but also favours those classes of applicant who have the time and, more importantly, the money to be able to afford to wait. Indeed, it is claimed that the high interest charges on land awaiting development permission is a fundamental factor in explaining the high rate of bankruptcies among small builders.

Analysing development control data

Planning applications represent a wealth of data about the nature and scale of development stretching back over more than 30 years. Ideally they should be used to provide information to help in deciding future planning decisions, but although their potential was realised by early planning thinkers like Patrick Geddes, it was not until many years after 1947 that they began to be effectively exploited (Geddes 1949). In 1962 the Ministry started to publish annual summaries of the planning applications determined by local authorities, but these were somewhat rudimentary and divided the applications into two basic categories, following the GDO and the Use Classes Order divisions – *(A) Building, engineering and other operations* and *(B) Change of use.* Each category was further subdivided into 8 smaller groups. In every instance the total number of applications was shown and the ratio of permissions to refusals. There were also further composite tabulations based on the different types of administrative area in England and Wales (counties, county boroughs) with a separate category for National Parks. Over the years the format has been expanded and there are now also tables for each of the standard regions, as well as statistics for appeals.

To compile even this amount of data on a national basis is a huge undertaking but, unfortunately, the statistics have serious shortcomings as a source of information about the national pattern of development pressures. For present purposes the overwhelming weakness is that no distinction was made between urban and rural areas. The only differentiation was between the pre-1974 Administrative Counties and County Boroughs, yet most of the former contained significant urban populations, for example Cheltenham in the County of Gloucestershire. There are also even more fundamental weaknesses. The statistics did not distinguish between the various types of application, so that there was no way of knowing what proportions were for outline, detailed, or detailed after outline permission. This, plus the fact that several applications were frequently submitted for the same site before one was successful, led to serious double counting for which it was impossible to allow accurately in any analysis. Nor are the applications weighted in any way, so that an application for 200 dwellings was treated in just the same way as one for a single house.

Gradually local authorities have begun to appreciate the potential of the development control statistics and, especially since 1974, they have started to set up monitoring systems on their own account. The main impetus for these new initiatives was not so much reorganisation itself as the replacement of Development Plans by Structure and Local Plans, both of which lay much greater emphasis on continuous review. Further encouragement was provided by a 1974 request from the DOE requiring more information on land use change from local authorities (DOE 1974e).

Application details

		USE CODE				type of premises (adverts)	floorspace	units	site area	existing land use	allocated land use	road class
		main		sub								
6	1											
6	2											
6	3											
6	4											
6	5											
6	6											
6	7											

Figure 5.8 Cornwall County Council planning application record. (Potter, D. 1975. Development control information. *The Planner* **61**, 110.)

Potter has described how a typical system was developed in Cornwall (Potter 1975) where the County Planning Office collects data from the six district councils to compile a planning process form for each application and also a planning application record (Fig. 5.8). The first simply gives the basic details of the application, while the second records the nature of the proposed development. The most important feature of this system and most others developed since 1974, is that the information is held on computer files so that it can be easily stored, manipulated and updated. Even so there are still significant shortcomings from a national point of view. Not all authorities have such a system and many of these are not mutually compatible, so that data cannot be compared. In any case most of the systems have not been operating long enough for them to have provided much significant feed-back into planning. In only a few instances has it proved possible to include data from before 1974 and, as a result, little is known still about the detailed variations in patterns of development control in England and Wales prior to that date.

One of the few exceptions has been the Planning Department of Devon County Council that from 1964 onwards produced a summary card (Fig. B.1) for every planning application in addition to the statutory register of planning applications. The information on each card is summarised in Figure 5.9, and can be divided broadly into 7 different groups:

(a) The location and nature of the proposed development and the name and address of the applicant.
(b) The type of application – outline, detail, detail after outline, or temporary.
(c) The implications of the proposal for policies in the County Development Plan, such as National Parks and settlement policies.
(d) The implications, if any, for those parts of the road system controlled by the Ministry of Transport.

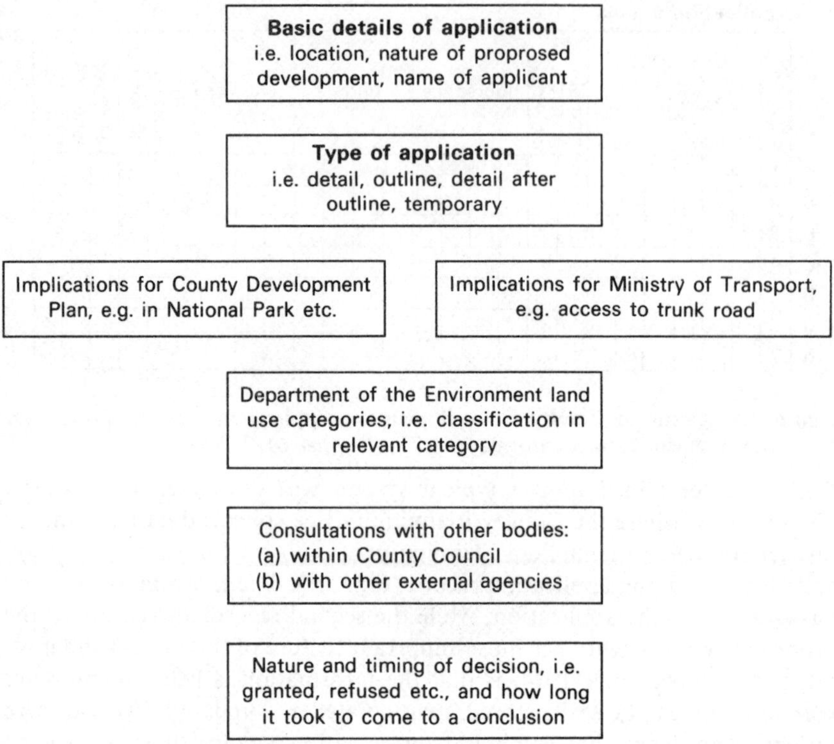

Basic details of application i.e. location, nature of proprosed development, name of applicant

Type of application i.e. detail, outline, detail after outline, temporary

Implications for County Development Plan, e.g. in National Park etc.	Implications for Ministry of Transport, e.g. access to trunk road

Department of the Environment land use categories, i.e. classification in relevant category

Consultations with other bodies: (a) within County Council (b) with other external agencies

Nature and timing of decision, i.e. granted, refused etc., and how long it took to come to a conclusion

Figure 5.9 Summary of planning application information stored on punch cards by Devon County Council.

(e) A classification of the development into one of the Department of the Environment standard land use, or land use change categories.

(f) Details of consultations, both with other county council departments and with outside agencies, such as the Countryside Commission and the River Authority.

(g) The nature and timing of the final decision – whether the application was granted or refused, whether conditions were attached, and the length of time it took to determine.

Even in this abbreviated form the data are unwieldy and difficult to assess. However, the authors have developed a system for using the information to measure the effect of planning controls on settlement change (App. B). There were a number of problems to be overcome. The first stemmed from the wide definition of development in the GDO that includes anything from a minor modification to the exterior of a building, to a proposal for a housing estate. In all cases the form of application is the same. The computer-based system classified the applications by type of development, thus distinguishing between major and minor proposals, allowing attention to be focussed on the more important items. This reduced the volume of data by more than half in all four case study areas. The second problem to be eliminated was that of double counting, which so severely limits the usefulness of the published statistics. As all the applications were identified by a grid reference accurate to 100 m, as well as by type of development, it was relatively easy to establish where there had been more than one application for a site and whether it was for outline, detailed, or temporary permission. The system was also able to allow for a third problem, the variation in size of applications. If development control data are to be used to monitor development, it is essential to know how many units are being applied for or what area they cover. Once the data were on computer file, they could be weighted in such a way that the subsequent analysis reflected the scale of any application. In other words an application for 20 houses was treated in the same way as 20 applications for individual houses. All the author's data presented in this book are weighted in this way.

Despite these improvements, there are still limitations to the usefulness of the data and any analysis needs to be undertaken with caution. For example, development exempted from planning control under the GDO is not included, so the overall picture remains incomplete. There is also the problem that even when a detailed application is granted, there is still no guarantee that the building will actually be built, even though this can be checked from building inspection records, or a field survey. These difficulties in interpreting development control statistics, even in those local authority areas where they have been overcome, have severely limited the amount of detailed analysis undertaken so far. Nonetheless, the few studies that have been completed illustrate the considerable potential of these data in explaining the effect of planning on settlement change.

Case studies of development control

Undoubtedly the pioneer work was David Gregory's 1970 book *Green Belts and development countrol*, and it has set the standard for most subsequent research (Gregory 1970). Gregory analysed all the planning application files lodged with the Seisdon Rural District in the West Midlands Green Belt between 1957 and 1966. His work incorporated a number of useful innovations, the most important of which was the decision to weight all applications by the area involved, thus creating a much more accurate picture of the spatial impact. The results showed that the true contribution of residential applications to the total in terms of area applied for is nearly 80%, whereas numerically they only account for just over 50% of the total. The study also distinguished between major and minor proposals and eliminated from the inquiry all applications for small and relatively insignificant developments, such as modifications to existing buildings, conversions and requests for garages and extensions.

The main purpose of the work was to relate planning permission to the area of land released for development and it showed that of the 3000 ha for which permission for major developments was sought in the West Midland Green Belt over the decade in question, 2596 hectares (86·5%) were refused by the local authority, although the proportion was subsequently reduced to 82·9% as a result of successful appeals to the Minister. The figures for residential development show an initial refusal rate of 97·7%, which was only reduced by a mere 0·3% on appeal. Gregory concludes that the local authorities adhered strictly to their stated policy of discouraging development in the Green Belt in the face of intense pressure and that they were strongly backed up by central government, particularly in respect of residential applications. Had the statistics been unweighted by area, and had minor applications been included, these conclusions would have been very much less convincing and it would have appeared that much more development had been permitted than was in fact the case.

Fortunately Gregory's preliminary work has been carried on in a wider study of residential development and land availability in the West Midlands by the Joint Unit for Research on the Urban Environment at the University of Aston. It was commissioned by the Department of the Environment and Gregory himself was a senior member of the research team. The results were presented in such a way that they were compatible with his earlier work. The study is not primarily concerned with development control, seeking rather to provide a case study of the problems facing house builders in the early 1970s, in view of rapidly rising property and land prices, an unprecedented rise in the number of planning applications, and allegations that local authorities were releasing insufficient land for residential development (Joint Unit for Research 1974, 1977). Planning records were but one of the primary data sources. The investigation covered the period from 1968/69 to 1972/73 and was by no means exclusively concerned with rural land. Nevertheless, it highlights many of the pressures facing the urban fringe. The area investigated was a corridor of land, extending from the centre of Birmingham south-westwards through the Green Belt to the zone beyond and included the pre-1974 rural districts of Kidderminster and Bromsgrove as well as 7 urban districts (Fig. 5.10).

Once again applications were weighted by the area applied for and it emerged that only 28·4% were on land zoned for housing and 46·6% were actually in the Green Belt, indicating that either applicants were blissfully unaware of the existence of the strategic policies in this part of the West Midlands, or that they believed that applications in supposedly restricted areas had a good chance of success. However, a comparison of success rates on different types of development land shows clearly that any belief in the latter was misplaced. The figures in Table 5.4 show that 61·7% of the land on which planning permission for housing was granted was in areas zoned for residential use and only 11·8% was in the Green Belt. Of the applications for housing in those areas zoned for residential development

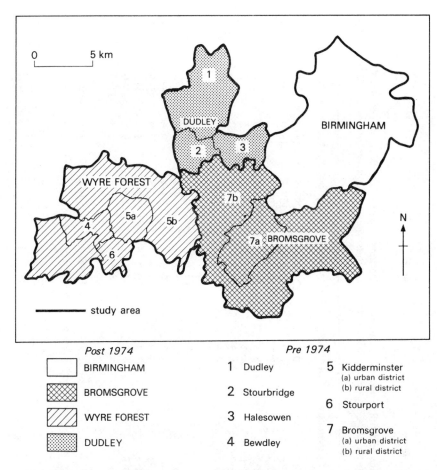

Figure 5.10 Local authority administrative areas in a part of the West Midlands before and after local government reorganisation in 1974 (*op. cit.*, Fig. 5.6).

53·3% were successful, but the proportions fell to 6·2% and 11·1% respectively in the Green Belt and Areas of Outstanding Natural Beauty. The analysis clearly shows how very useful weighted development control statistics can be for evaluating the operation of strategic policies, since they give a much truer picture of the way in which policies have been interpreted, than the raw data does.

The data also revealed other interesting aspects of the planning system. There is, for instance, considerable variation from year to year in both the quantity of land applied for and the amount released. In 1969/70 only 28·3% of the residential acreage applied for on land zoned for housing was granted, whereas in 1970/71 the comparable figure was 62%. At the other end of the spectrum, Green Belt success rates fluctuated annually between 1% and 13·6% as shown in Table 5.4. Despite these annual variations, the grounds for granting or refusing are very consistent. In the rural districts,

Table 5.4 Gross area of land, the subject of outline and full residential planning applications in part of the West Midlands 1968–73 (ha). (Joint Unit for Research into the Urban Environment 1974. *Land availability and the residential land conversion process*. Birmingham: University of Aston.)

	Housing	Industry and Commerce	Open space	White land	Green belt	AONB	Others	Total
(a) land applied for								
1968–69	103·4	0·2	10·7	121·2	299·5	15·3	23·6	573·9
1969–70	130·2	0·3	20·9	2·8	157·8	13·0	6·3	331·3
1970–71	130·8	0·3	19·5	10·1	102·7	14·8	35·1	313·4
1971–72	110·2	2·1	23·4	39·6	215·5	13·6	19·0	424·1
1972–73	146·4	7·8	21·8	90·1	242·8	31·5	3·6	544·0
total	621·0	10·7	96·3	263·8	1018·3	88·2	87·6	2186·7
%	28·4	0·5	4·4	12·1	46·6	4·0	4·0	100·0
(b) land released								
1968–69	64·9	—	5·9	23·6	3·5	3·3	6·9	108·1
1969–70	36·8	0·2	2·2	5·8	4·7	3·6	6·4	59·7
1970–71	81·1	—	8·3	0·3	11·4	0·6	18·7	120·4
1971–72	63·5	0·1	3·2	7·2	10·8	0·9	28·9	114·6
1972–73	84·7	0·5	2·3	10·7	32·9	1·4	1·1	133·6
total	331·0	0·8	21·9	47·6	63·3	9·8	62·0	536·4
%	61·7	0·1	4·1	8·9	11·8	1·8	11·6	100·0
(c) success rate								
% success rate	53·3	7·5	22·7	18·0	6·2	11·1	70·8	24·5

where least land is zoned for development, the main reasons for refusal can be classed as deviations from the published Development Plan policies, but in the urban districts, where most of the land is available, site character and the nature of the development itself are much more prominent reasons for refusal. The analysis went even further and examined the applicants, as well as the reasons for refusal, revealing a marked difference between rural and urban areas. Far more private individuals apply in the country and in general they appear to be less well-informed about published strategic policies, than the corporate applicants who predominate in the towns. While this may be hardly surprising, the relative lack of success amongst private individuals, for whatever reason, must be a source of discontent and disillusionment with the way in which the planning system works.

Further examples are provided by the authors' work in Devon which was chiefly concerned with illustrating how strategic policies, particularly those concerned with landscape conservation, had been implemented through the development control system (see Ch. 6). A major difference from the West

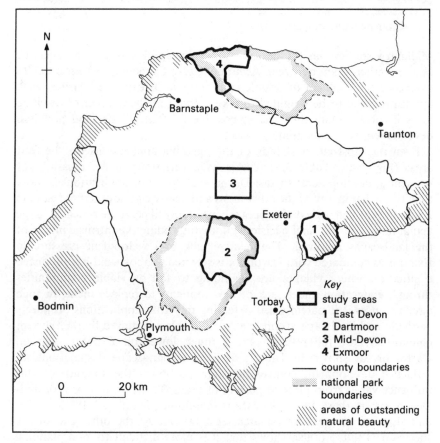

Figure 5.11 The four Devon field study areas.

Table 5.5 Planning applications for residential development in the four study areas 1964–73. (Authors' survey.)

	East Devon	South-east Dartmoor	Mid-Devon	Western Exmoor
1964	119e	72	33	15
1965	119e	94	46	10
1966	119e	81	31	14
1967	107	45	38	6
1968	90	48	40	13e
1969	62	57	13	13e
1970	95	55	28	13e
1971	146	67	33e	18
1972	216	106	33e	33
1973	119e	99	33e	15
total	1073	724	328	150
area (km²)	135	231	99	145
density of applications per km²	7·95	3·13	3·31	1·04

e = Estimate based on partial data.

Midland's studies was the decision to weight applications by numbers of houses rather than by area. Accurate grid-referencing was essential to overcome the problem of double counting and as explained earlier, each application had to be weighted to take account of the number of dwelling units involved. Only then was it possible to relate actual and potential development to the planning process.

Even the most cursory study of the planning applications for the four areas (Fig. 5.11, Table 5.5) reveals wide variations in the pressure for residential development. In east Devon the density of applications is more than twice that in any of the other areas and nearly eight times that found in Exmoor and such wide differences are inevitably likely to be reflected in the actual nature of the final decisions, notwithstanding the intrinsic merits of the landscapes involved. The relative lack of development pressure in Exmoor as compared with Dartmoor is referred to above and has produced a situation where villages are beginning to fail as viable communities through emigration and a lack of new residents to replace them. In such circumstances it is patently self-defeating to refuse applications on purely aesthetic and landscape grounds. Clearly this is recognised by the relevant committees, for the refusal rate is much lower on Exmoor than on Dartmoor, as shown in Table 5.6. Even so landscape designations do influence development control decisions, although the strength of this influence varies from place to place. For example, the refusal rate for mid-Devon, where there are no landscape inhibitions, is only 20·1% less than half that for the next lowest area, east Devon. At the other end of the spectrum, Table 5.6 also shows that it is more difficult to gain planning permission in either of the National Parks than elsewhere in the county.

Table 5.6 Decisions on development control applications in the four study areas 1964–73. (Authors' survey.)

	Approved: no conditions		Approved: with conditions		Refused	
	no.	%	no.	%	no.	%
East Devon	138	12·9	502	46·8	433	40·3
South-east Dartmoor	42	5·8	286	39·5	396	54·7
Mid-Devon	22	6·7	240	73·2	66	20·1
Western Exmoor	5	3·3	82	54·7	63	42·0
total	207	9·1	1110	48·8	958	42·1

Table 5.7 Decisions on development control applications by category of application in the four study areas 1964–73 (%). (Authors' survey.)

	Approved: no conditions	Approved: with conditions	Refused	Total
East Devon				
outline	0·4	22·2	77·4	100·0
detail	36·4	53·3	10·3	100·0
detail after outline	8·1	10·3	5·9	100·0
South-east Dartmoor				
outline	—	25·9	74·1	100·0
detail	23·9	54·0	22·1	100·0
detail after outline	10·4	78·3	11·3	100·0
Mid-Devon				
outline	1·0	62·1	36·9	100·0
detail	—	85·7	14·3	100·0
detail after outline	14·5	85·5	—	100·0
Western Exmoor				
outline	1·2	41·7	57·1	100·0
detail	13·0	78·2	8·8	100·0
detail after outline	—	100·0	—	100·0

In many respects, whether or not an application is granted or refused is less important than the safeguards that are attached to any permission. Table 5.6 is encouraging in this respect, since it reveals that only 9·1% of all the applications lodged in the four areas are granted unconditionally. Conditions may relate to any aspect of the application and can be as much concerned with landscaping and layout as with the details of building design. As can be seen from Table 5.7, conditions are just as likely to be attached at the outline stage, where the general principle of the application and questions of overall layout are determined, as at the detailed stage, where the design of the buildings themselves are decided. Typical conditions at the outline stage specify items that will need to be included in the subsequent detailed application. An example of such conditions are the following, which were attached to an outline permission for a dwelling at Yelverton in the Dartmoor National Park:

' (a) The development hereby authorised shall be carried out only in accordance with detailed drawings which shall previously have been submitted to and approved by the Local Planning Authority. Such detailed drawings shall show the design and external appearance of all proposed buildings, their siting, the materials of which they are to be constructed, the arrangements for the disposal of foul and surface water, the means of access from public highways, areas for vehicle parking and all other works including walls, fences and other means of enclosure and screening. The drawings shall indicate the location and species of all trees existing on the site.

(b) A garage or hardstanding and parking space for motor vehicles shall be provided and sited in accordance with details which previously shall have been submitted to and approved by the Local Planning Authority and the dwelling shall not be occupied until these facilities and vehicular access thereto have been provided. These facilities shall be kept permanently available for the parking of motor vehicles.

(c) Before the occupation of the development hereby permitted, a new section of hedgebank shall be constructed, in accordance with details, which shall previously have been submitted to and approved by the Local Planning Authority, to close the existing gap in the hedgebank on the boundary fronting onto the A 386, and no vehicular or pedestrian access shall be permitted direct onto the A 386. '

When conditions are attached at the detailed application stage they are normally much more specific and relate to materials and actual modes of construction. Again an example, in this case from the village of Holne in the Dartmoor National Park, will serve to illustrate the point:

' (a) The roofs of the new building and the stone outbuilding for staff accommodation shall both be in natural slate, samples of which shall previously have been submitted to and approved by the Local Planning Authority.

(b) The new gable indicated on the outbuilding conversion drawings shall be omitted to the satisfaction of the Local Planning Authority.

(c) The proposed trout pools shall be lined in an appropriate material and the lining maintained to the satisfaction of the Local Planning Authority to prevent any loss of water by soaking away and spillage. '

Conditions such as these are an essential part of the planning process, for it is through them, just as much as the power to grant or refuse that a planning authority is able to control the nature and extent of development. It is, however, one of the great ironies of the planning system, that while an authority can insist on quite small details of design, colour and construction materials when the building is proposed, many of these items can subsequently be changed once the project is complete without any reference to the planning authority.

Despite the obvious objections to a two-stage planning application in terms of administrative time and money, the outline stage is still a very useful screening device, allowing an applicant to find out relatively easily whether a full application is likely to be successful. Table 5.7 shows indeed that the bulk of refusals are at the outline stage and that only a very small proportion of detailed applications, submitted after outline permission has been granted, are then turned down. The opportunity to be able to test the principle of development, before proceeding to detailed matters of design, is particularly useful in sensitive rural landscapes like National Parks where there is a strong presumption against development. This point emerges clearly from the figures in Table 5.7. The chances of an application being refused in mid-Devon (36·9%) are strikingly lower than in the other three areas where the landscape policies mean that development is generally discouraged.

The statistics from the two National Park areas in the study exhibit some interesting contrasts. It has already been mentioned that though there is a strong commitment both nationally and locally to preserving Exmoor's scenery, the general decline of rural communities in north Devon make it difficult to do anything other than encourage a developer who wishes to invest in new housing. In eastern Dartmoor the situation is quite different, with strong pressure for development in all the villages but little official encouragement, except in the two key villages. The difference in the relative locations of the two National Parks is thus the prime reason why the proportion of outline applications refused in the Dartmoor study area is 74·1%, while on Exmoor the figure drops to 57·1%.

Development control data can also be used to show how the nature of decisions varies over time as illustrated in Table 5.8, which shows that refusal rates varied quite considerably from year to year, both between the four areas studied and within them. However, only in east Devon are there sufficient applications in any one year for a clear pattern to emerge. In this case there is a strong relationship between the refusal rate and the total number of applications. As the number of applications grows, so the proportion of refusals falls. Only the most tentative suggestions may be made about the causes of this relationship but it does seem possible that precedent may well play a part. Planning committees are always strongly influenced by previous decisions, especially those from the recent past. In a year with a large number of applications, one permission will lead almost

Table 5.8 Refusal rates for residential planning applications in the four study areas 1964–73 (%). (Authors' survey.)

	East Devon	Dartmoor	Mid-Devon	Exmoor
1964				
1965		45	20	20
1966		53	23	57
1967	42	44	26	17
1968	50	27	13	31
1969	47	56	15	23
1970	39	33	11	31
1971	37	48		39
1972	24	56		48
1973		64		
	$\bar{x} = 41 \cdot 50$	$\bar{x} = 47 \cdot 33$	$\bar{x} = 18 \cdot 00$	$\bar{x} = 33 \cdot 25$
	$\sigma = 5 \cdot 50$	$\sigma = 15 \cdot 92$	$\sigma = 5 \cdot 88$	$\sigma = 13 \cdot 05$

inevitably to others on the grounds of comparability alone and there is some evidence that this happened in the case of east Devon in the early 1970s. It should perhaps be noted that these findings contrast with the national figures that reveal a very close relationship between the rate of refusals and permissions, as shown in Figure 5.12. This correspondence is quite striking and leads one to conclude that, in general, local authority development control staff are determined to keep the rate of refusal constant, whatever the application rate may be, notwithstanding the particular evidence from east Devon.

The above studies are all independent analyses of the ways in which development control systems have worked, but a number of local authorities have also produced studies of specific aspects of the planning process based upon development control records. For example, an analysis by Cornwall County Council as part of the preparatory work for their Structure Plan, discussed the way in which development control had been carried out in the county's Areas of Outstanding Natural Beauty and is of particular interest (Cornwall County Council 1976). Once again it showed that the majority of applications in these areas were for residential building and concluded that this was the only category of development under planning control of general and widespread significance for rural landscape change. Since the Cornish Areas of Outstanding Natural Beauty were designated in 1959, there have been some striking changes in the actual pattern of decisions. Although planning permission generally has become progressively easier to obtain in the county as a whole, in the Areas of Outstanding Natural Beauty the refusal rate has risen from an average of 20% for the years 1959–64 to 32% for the years 1969–74. Prior to 1959, 56% of permissions were also issued unconditionally, but since then the proportion had fallen dramatically to only 3%, mainly as a result of a more general environmental awareness. Nevertheless, in the opinion of the planning authority a great deal more could still be achieved. Only a

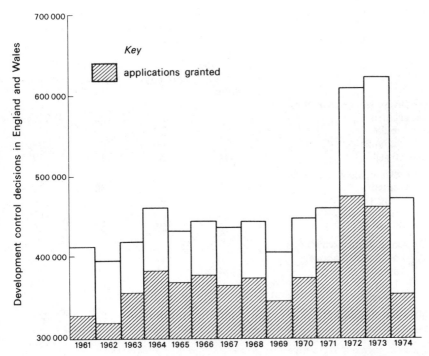

Figure 5.12   Planning decision statistics. (House of Commons Paper 564 1977–8.)

minority (13%) of the conditions, related to such things as siting, design, landscaping and the materials to be used, the majority were concerned with access and other highway considerations. Indeed, the authority concluded that there was considerable cause for concern about the way in which the development control system had operated in these particular parts of Cornwall and the report stated that, 'the policy set out in the approved County Development Plan for the protection and enhancement of Areas of Outstanding Natural Beauty has apparently been too broadly worded to give a clear cut lead on the protection of large scenically attractive areas of Cornwall' (Cornwall County Council 1976, p. 11). The dangers are emphasised by the fact that the number of applications submitted in the Areas of Outstanding Natural Beauty was much greater than in the county as a whole and grew six-fold between 1954 and 1974, nearly double the rate of increase in the rest of the county over the same period. Thus, although the refusal rate has increased over the years, the number of successful applications has been growing even faster, producing a substantial amount of development in these attractive rural areas.

A rather different kind of study was undertaken by the planning department of the pre-1974 Welsh county of Denbighshire. The study was concerned with house construction in Ruthin Rural District and tried to show how the pattern had changed in the 10 years between 1963 and 1972.

Table 5.9 Planning applications (and number of dwellings) for rural housing, Ruthin RD, January 1963–December 1972. (Denbighshire County Council 1974. *Rural housing in Denbighshire.* Ruthin: the Council.)

	1963–72		1968–72	
	Applica-tions	Dwellings	Applica-tions	Dwellings
(1) Inside village envelopes				
granted	53	117	52	284
built		(85)		(75)
unbuilt		(32)		(209)
refused	14	26	16	169
(2) Outside village envelopes				
granted	38	53	38	42
built		(46)		(29)
unbuilt		(7)		(13)
refused	56	389	67	466
(3) Agricultural permissions (not included above)				
granted	34	34	29	29
built		(19)		(18)
unbuilt		(15)		(11)
(4) Total Ruthin RD	195	619	202	990

The level of applications had remained fairly stable but there had been an increase of more than 50% in the number of dwelling units applied for (Table 5.9), confirming the necessity for weighting development control data so as to take account of the number of units applied for. The records vouched for the overall success of the strategic settlement policy that sought to restrict new residential development to existing villages. Refusal rates were much higher outside the settlements than within and the policy had been enforced progressively more rigidly over the time period.

The Ruthin findings, although they broadly agree with the results of the other studies discussed, also provide some interesting new insights. First, the existence of the policy trying to discourage new development in the open countryside seems in no way to have stopped people from applying for planning permission outside the existing settlements. The bulk of applications were in areas not zoned for development, confirming the conclusion from all the other examples considered that there was considerable ignorance of planning policies on the part of many applicants. Secondly, by separating out agricultural permissions the figures give a more realistic picture of the efficacy of the settlement policy. Indeed agricultural permissions were found to account for roughly a quarter of the total and, therefore, made a considerable contribution to new development in the countryside. Thirdly, as shown in Table 5.9, a high number of permissions remained unbuilt. This finding would seem to be rather different from Devon, where in the four areas studied 75% of permissions resulted in new buildings. If buildings are not started once planning permission has been granted, the planning authority is put in a very awkward position for they are unable to assess new applications against those already granted. This matter has been studied further by the Economist Intelligence Unit. Their report, *Land availability. A study of land with residential planning permission*, to the DOE and the Housing Research Foundation found that by 1977 some 83% of the mid-1975 stock of permitted sites had either already been developed, or were currently being developed, or were likely to be developed in the near future. Two main reasons were cited for failure to build: planning difficulties and a downturn in the market. Interestingly, 60% of the land where development had not commenced was owned by non-house builders, suggesting a high degree of land speculation. The study concluded that this was the most likely explanation of delay. Overall the results indicated that planning permissions are 70% accurate as an indicator of 'real availability' of housing land, if real availability is taken to mean that a start may be made on the site within 2 years (Economist Intelligence Unit 1978).

The disparate nature and lack of co-ordination between these case studies has already been discussed, but, despite this, some common threads do run through them all and it is worth drawing them out at this juncture. The most important feature is the overwhelming predominance of new residential building in all the areas. Not only is it the largest category numerically, it is also the one that rural planning policies are most frequently designed to control. Therefore, there seems to be ample justification for concentrating the analysis on this category of development. As for the policies themselves, the results indicated that while planning authorities generally adhere to them, they have little direct influence in helping to guide intending applicants, especially private individuals, to those areas designated for development. A considerable amount of official time is spent processing applications that have little chance of success, simply because they are contrary to published policies. Indeed this is one of the strongest arguments for retaining the two-stage application process. The Devon study showed how effective the outline stage is in acting as a filter for screening unrealistic applications. The evidence for more stringent standards of development both in protected landscapes and elsewhere in the countryside is inconclusive. Conditions are attached to all but a handful of permissions at both the outline and detailed stages but frequently these apply only to engineering and highway considerations, rather than to the equally important questions of design, landscaping and aesthetic impact of the new development.

The development control system assessed

It is a tribute to the 1947 planning system, that it survived so long without

major reappraisal or modification. The method of plan making was not
seriously questioned until the mid-1960s and is only now, 10 years later,
actually in the process of reform. Development control, which initially was
only intended to provide back up powers for the Development Plans of
local authorities, has undergone little administrative change, although it has
come to play an increasingly important part in the planning process.

In the mid-1970s, however, a number of events began to undermine the
development control system. Local government reorganisation in 1974 put
responsibility for development control in the hands of the districts, rather
than the counties that retained the strategic planning functions. Adjusting
to the new procedures caused many problems in itself and the difficulties
were compounded by an 'unprecedented and unforseeable increase in the
number of planning applications to local authorities and appeals' (DOE
1975a, p. 3). From 414 300 applications in 1970 the number rose to 622 700
in 1973, an increase of over 50%. Such growth severely strained the whole
planning system and led in turn to long and frustrating delays for
applicants, particularly in the case of large and complex proposals requiring
more than routine consideration. At the same time there was a strong
demand for greater direct public participation and, laudable as this may be,
it further added to the time taken to determine applications. Faced with all
these different pressures, the government had little option but to review the
working of the development control system.

The government's chief concern were the charges of apparently needless
delay and bureaucracy involved in the processing of even the simplest
application and, in 1973, it set up an inquiry under George Dobry QC to
look into all aspects of the development control system. The terms of
reference were:

 ‘ (a) To consider whether the development control system under the
 Town and Country Planning Acts adequately meets current needs and to
 advise on the lines along which it might be improved, bearing in mind
 the forthcoming redistribution of planning functions between local
 authorities and the new system of Structure and Local Plans;
 (b) To review the arrangements for appeals to the Secretary of State
 under the Planning Acts, including the rights of appeal and the handling
 of appeals in the Department of the Environment and to make
 recommendations ’ (DOE 1975a).

Dobry's final report was published in 1975 and was based on five clear and
straight-forward principles:

 ‘ (a) Giving greater freedom to harmless development, but
 (b) guarding against harmful development by retaining applications
 for all cases, as at present;

(c) separating from the main stream all applications which might cause harm;

(d) disposing of applications in the mainstream by rapid and routine procedures; and

(e) applying the same approach to appeals ' (DOE 1975a, p. 6).

To these ends he made a large number of recommendations, the most important of which was to divide applications into (A) and (B) categories, corresponding with minor and major development proposals respectively. The former would be dealt with quickly and routinely, the latter would receive much more careful scrutiny. He also recommended that there should be better information for the public about what developments required planning permission, how an application should be made, and what rights of appeal there were if an application were rejected. The report also made a number of suggestions as to how the committee structure could be made to work more efficiently.

As far as the countryside was concerned the most important part of Dobry's report is Chapter 9, which examined ways in which the GDO might be relaxed, and what special safeguards needed to be built in for those areas Dobry termed 'Special Environmental Areas'. Any relaxation of the GDO was rejected on the grounds that amenity would be endangered and there was a suggestion, albeit a somewhat reluctant one, that there should be tighter controls over certain types of agricultural building. The reluctance stemmed from the fact that further controls might inhibit the farming industry from achieving the government's stated aim of increasing the domestically produced proportion of the nation's food (Cmnd 6020 1975). Despite such reservations, Dobry recommended that planning control ought to be extended to certain categories of agricultural building, which had previously been exempt:

(a) Buildings within 100 m of existing buildings,
(b) Intensive building groups, which although for the use of livestock are not requisite for the use of land for agriculture.

In the Special Environmental Areas, which were divided into those of national importance (National Parks, Areas of Outstanding Natural Beauty, the proposed Heritage Coasts and the more important conservation areas) and those of local importance (other conservation areas and other areas of town and country identified in Development Plan policies), Dobry examined the case for special new measures. In the areas of national importance he recommended that the Landscape Areas Special Development Order should be much more widely employed than at present. In the areas of local importance he felt that everything that needed doing could be achieved by Article IV directions, which bring into planning control those uses that are normally excluded by the GDO. The government's reaction to the reports was to welcome them warmly, recommend that they be widely read by students of planning and to reject outright, or delay action on all the main proposals (DOE 1975b). The division of planning applications was deemed to be unworkable from a practical point of view because it would increase delays in determining applications. The government's view was that the development control system was fundamentally sound and that almost everything that needed to be done could be achieved within the compass of existing legislation. Better planning management was what was required. They also argued that the crisis caused by the growth in the number of applications had passed and there certainly has been a decline in the number of applications since 1973 (Fig. 5.12). As local government has begun to settle down after reorganisation, the planning system is dealing with applications much more expeditiously as has been shown earlier in this chapter (Storey 1978).

The response to the specific proposals relating to agricultural buildings and Special Environmental Areas was cautious and non-committal. The government pointed out that the GDO is in any case subject to periodic review and promised, at an appropriate time, to consider reducing the range of agricultural buildings excluded from planning control and either extending the Landscape Special Development Order, or introducing some other form of Special Development Order. Ironically, when the General Development Order was eventually reviewed in the summer of 1978, it proposed not an extension but a relaxation of controls, but the ensuing outcry from the planning profession and amenity lobby caused the proposals to be withdrawn after a few months' discussion in December 1978. Similar proposals were nonetheless resurrected in March 1980 in a draft GDO. This proposed a doubling of the exceptions for most minor developments, except in National Parks, Areas of Outstanding Natural Beauty, and Conservation areas. Even more radical reforms were suggested in August 1980 in a draft circular from the DOE. This proposed that, regardless of zoning, there should be a presumption in favour of granting planning permissions except in green belts, National Parks and on good quality agricultural land. The circular also advocated the relaxation of controls over aesthetics and design (DOE 1980).

Another study of development control procedures has been conducted by the Environment Sub-Committee of the House of Commons Expenditure Committee, but it is very different to Dobry. It was initiated to examine the widespread complaints about delays and the high cost of the planning process. Its findings were based on a wealth of evidence given to the Sub-Committee by numerous official and private agencies. Its report is one of the most comprehensive commentaries on the planning process ever produced, but the evidence, running to over 800 pages, is almost inevitably contradictory (Expenditure Committee 1977). Developers argued that there were excessive delays in determining applications and that this was unnecessarily forcing up their costs and in consequence house prices. Conversely, most amenity societies believed that the present system was too rapid and did not give sufficient time for damaging developments to be opposed. The evidence from both central and local government was that the system was working remarkably well and that there had been great improvements in the development control system since local government reorganisation. A view typical of the official line is provided by Vickery, 'But the greatest change, perhaps resulting from the increased status and resources that development control now commands, is that it is becoming a positive influence in planning, rather than the totally negative control practised formerly. Dobry's change of attitude towards a more positive and constructive approach has crept up on the planning authorities despite Government advice rather than because of it' (Vickery 1978, p. 25). In other words the present development control system has successfully accommodated the changed circumstances in which it finds itself.

The Sub-Committee took the view, however, that all was not well and made two main recommendations. First, it proposed that a team of planning assessors be appointed to scrutinise the performance of local authorities and to help those experiencing difficulty, with the general aim of raising all to the level of the best. Secondly, and more radically, it proposed that there should be a complete review of the plan-making system. The Sub-Committee argued that the hierarchy of Structure and Local Plans was not working out in practice and that in the absence of proper guidelines the whole process of development control was breaking down. Davies has put the case succinctly, 'Developers, particularly, felt that the absence of up-to-date authoritative development plans created uncertainty, contributing to delays in development control. Other witnesses reflected on the complexities and uncertainties of plan-making, notably in the relationship between regional and structure planning, and the Committee were particularly concerned about the very slow, and apparently uncertain progress in the preparation and approval of both structure and local plans' (Davies 1978, p. 147). The lack of coincidence between plans and development control decisions has often been asserted but in the absence of more detailed studies, such as those referred to earlier, it has been impossible to prove the case one way or the other. Certainly the Sub-Committee was not presented with any new empirical evidence enabling it to resolve the debate.

In its response to the report the government gave a cautious welcome to the idea of planning assessors, but made no firm commitment to the setting up of such a service (Cmnd 7056 1978). They rejected outright the proposed review of the plan-making system. The Sub-Committee was so concerned at this negative reaction that it reconvened and produced a second report, which reiterated its conclusions and asked the government to reconsider their position (House of Commons Paper 564, 1978).

The result of these many thousands of words, and long weighty deliberations is something of a stalemate. Doubts about the system persist, but they are least strong amongst the government and local authorities who operate it. In general they believe that things have improved substantially since 1974 and will continue to do so. Nevertheless, criticism and new proposals for reform are still appearing. The Royal Town Planning Institute has recently published a paper which proposes a two-tier system of applications, but dissimilar to that proposed by Dobry, and much clearer guidelines for both applicants and planners (Royal Town Planning Institute 1978).

In conclusion, development control remains something of an enigma, especially in rural areas, where there are persistent doubts about the extent to which it should or should not apply. The system itself is legislatively complex and something of a lawyers' paradise and, as such, remains a mystery to large sections of the general public. On the other hand, there are few people who do not at some time come into direct contact with planning, either as an applicant or as someone directly affected by a proposed development. That they are unable to understand easily why the process is necessary and what is actually happening, compounds the general feeling of suspicion and unease. The problem is that nobody fully understands what is happening and how the system is actually working, especially in rural areas where the controls are so much less comprehensive. The situation was pithily summarised by Dobry in evidence to the House of Commons Expenditure Sub-Committee when he said that, 'Mr Lloyd Thomas (Assistant Secretary, Department of the Environment) has not got the faintest idea what is happening in the country' (Expenditure Committee 1977, p. 1048). Nor, it should be added, in the absence of more detailed analyses, such as the case studies described in this chapter, has anybody else. The welter of words over the past 5 years on the subject of development control has really only served to expose the lack of factual information about what in fact it has achieved.

6 Settlement change

Changes in population structure and life-styles

The population of the United Kingdom rose from 38·2 million in 1901 to an estimated 56 million in 1976. The increase was accompanied by significant changes in the demographic structure and introduced important new elements into the rural settlement pattern. As Figure 6.1 shows, fluctuations in the rate of growth of the population roughly follow a 20 year cycle, each upturn coinciding with a peak in the demand for new housing in the countryside. This was underlined by the surge of planning applications in the early 1970s discussed in the previous chapter (Fig. 5.12). The surges in demand have not, however, been uniformly spread for migration has redistributed the British population not only generally southwards, but also into specific rural areas, as shown in Figure 6.2 (Weinstein 1975). The

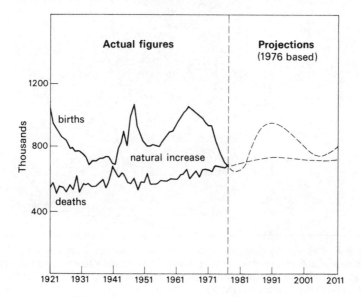

Figure 6.1 Population changes and projections for the United Kingdom 1921–2011. The erratic nature of the birth rate sends a series of surges through the age group structure of the population, which results in peaks of demand for new housing 20 to 25 years later. (Central Statistical Office 1977. *Social trends 1977.* London: HMSO.)

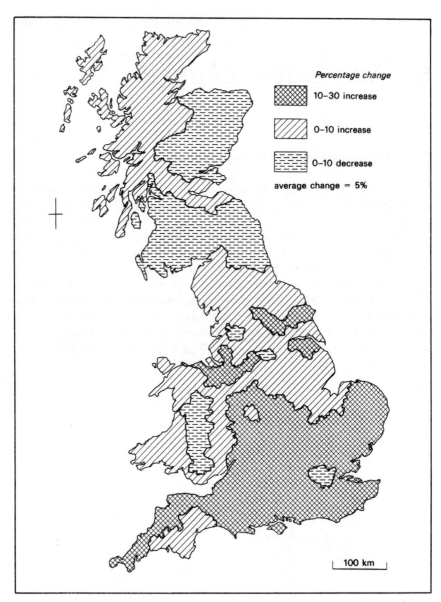

Figure 6.2 Population change by sub-divisions of standard regions 1961–71. The map demonstrates two types of rural Britain. The declining remote and upland areas and the expanding accessible and lowland areas. (Central Statistical Office 1974. *Social trends* 1974. London: HMSO.)

remoter upland areas of the north and west are actually faced with the problems of decline and the very survival of many rural settlements in these areas has been brought into question. Nevertheless, in rural Britain as a whole there was an increase in population of 1 698 000 between 1961 and

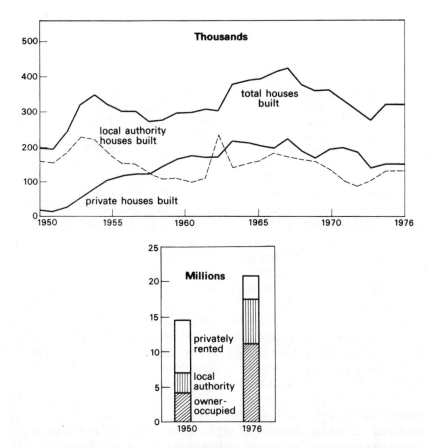

Figure 6.3 Houses completed and dwellings by tenure, in the United Kingdom, 1950–76. In spite of the original intention of the 1947 planning act that most new houses should be provided by local authorities, rising prosperity and a swing of political attitudes in favour of home ownership has led to a rising demand for new houses, which has been largely met in the countryside. (Central Statistical Office 1977. *Social trends 1977.* London: HMSO.)

1971, while the conurbations and other major urban areas suffered a loss of 884 000 people (Champion 1976).

It is difficult to assess precisely how many houses have been built in rural as opposed to urban areas since the war. But as most new dwellings have, by necessity, been constructed either in the countryside, or in the existing urban fringe where land is available, the national figures for house completions give a reasonable estimate of the numbers of new houses in rural areas. Between 1951 and 1976 the overall stock of dwellings rose from

Table 6.1   Household size 1951–71. (Central Statistical Office 1977. *Social trends 1977.* London: HMSO.)

Household by size %	1951	1961	1971
1 person	10·7	11·9	18·1
2 person	27·3	29·8	31·5
3 person	25·1	23·3	18·9
4 person	19·0	19·1	17·2
5 person	9·7	9·2	8·3
6 person	8·2	6·6	6·1

14·8 million to 20·6 million, with private construction predominating from the late 1950s onward (Fig. 6.3). A more precise analysis of where these houses have been built and how much rural land they have consumed is included in the next section of this chapter.

This still leaves unexplained the detailed reasons why people have tended to move to rural areas in such large numbers. Population growth on its own is not a sufficient explanation, especially since it has been accompanied by a rapid decline in household size (Table 6.1), which has meant that even in areas where the population has been stable, there has still been a demand for new houses. The reasons for the decline in household size are the decision of parents to have smaller families and the desire on the part of many young people to leave the parental home sooner than they used to *(Office of Population Censuses 1976).*

Whatever the reasons, the net effect has been a demand for more and smaller houses, which has been easiest to meet in rural areas. Ambrose has examined the whole question of urban–rural migration and proposed a convincing justification for the drift from town to country (Ambrose 1974). New houses are usually easier to buy because they attract the most favourable mortgage terms and, for the reasons of ease of land assembly and availability, the bulk are built on green field sites in the countryside around existing settlements. Lower land costs also mean that rural houses are usually cheaper than their urban counterparts and, in addition, rural rates have traditionally been lower than those in towns and cities (Green 1980). Finally, there is a widely held though often erroneous belief that the country is not only more friendly and sociable than the town, but also healthier. It is clear that for many people these attractions offset the disadvantages of long journeys to work and the relative isolation from shops, services and entertainment.

It is widely expected that the countryside will continue to exercise considerable attractions especially in the short- to medium-term even if, as predicted, the rate of population increase in the 15–25 year age group slows down (Fig. 6.1). More and more people view their house not merely as a shelter but as a medium for expressing their personality and the life-style to which they aspire. One of the privileges of the affluent society is that it allows home and place of work to be widely separated; a house in the country enables people to partake of many of the pleasures of country surrounded by urban comforts, yet, at the same time, shielded from the more unpleasant and mundane aspects of both environments.

The prospect of escapism that the countryside provides has meant that even in areas where the permanent population is static or declining, there is often a demand for new homes for retirement and holidays (Ashby *et al.* 1975). The desire for such accommodation exists in the areas of growth as well, but higher house prices there tend to deflect demand to places where the economy is less buoyant and the landscape less developed.

It should be made clear that all the above factors are only truly relevant to the private housing market and, although 55% of houses are now owned privately, rural areas still contain considerable numbers of both council houses and privately owned tenanted dwellings, in particular tied cottages, where different conditions apply. Commuting may be a feasible, even attractive, proposition for the professional worker who can afford two or more cars, but it is very expensive and inconvenient for the less well paid council tenant, relying on the erratic and sporadic transport facilities available in rural areas. The mismatch between jobs and housing has been one of the worst failures of post-war rural planning. The numbers of new council houses foreseen in most plans has outstripped the numbers of new jobs, indeed in many areas jobs have actually declined. Public transport has also suffered because services for small and scattered populations are almost inevitably uneconomic and this has created a vicious circle, in which car ownership becomes first desirable and then essential as loss of business accelerates the withdrawal of even the remaining services (Moseley 1979). The loss of public transport has led to serious rural deprivation in certain areas as the old, the young and the majority of housewives are denied the freedom offered by the private car (Shaw 1979). Even though 70% of rural households own a car, these particular groups are often more isolated than they were 100 years ago.

There has, therefore, been considerable pressure for new housing in rural areas since 1945. People have been drawn not by the prospect of employment, but by the availability of relatively cheap housing and the amenities of country life and these advantages have outweighed the relative lack of social, medical and transport services. On the other hand, for a substantial minority, mainly council tenants, there has been little choice and rural life for them means poor job prospects and isolation. This chapter now considers the effect that planning has had on these changes in general and especially on four aspects: urbanisation of the countryside, village settlement patterns, village layout and building design, and major development projects undertaken to service the growing numbers of people.

Urbanisation and the countryside

One of the overriding concerns of the interwar years was the apparently
uncontrollable growth of urban areas, in particular the Greater London
conurbation. Fears were repeatedly voiced that the countryside and
valuable agricultural land were gradually but inexorably being enveloped by
a tide of urban sprawl, and that living conditions within cities were so bad
that they seriously threatened the health of the nation's population. These
fears found their most cogent and influential expression in the Barlow
Report (Cmd 6153 1940), which recommended that efforts should be made
to spread the urban population of England and Wales more evenly among
the countries' towns and cities.

Weight was added to these arguments by the results of the somewhat
limited empirical studies of land use undertaken at the time. Ashby writing
in 1939 noted, 'the very interesting situation that in England and Wales
both Greater London and the rural districts recently have been gaining by
migration: London at the expense of other counties and other internal
areas, and the rural districts from county boroughs and probably somewhat
from other urban districts' (Ashby 1939, p. 356). This confirmed both the
increasing concentration of population in London and the expansion of
non-agricultural settlement in rural areas. Further corroboration came both
from the results of the First Land Utilisation Survey that were used by
Stamp to show how the economic depression in agriculture and low land
prices had led to serious losses of farmland to urban uses in the two decades
before the Second World War (Stamp 1962), and from the 1939 *Report on
the preservation of the countryside* (Health, Ministry of 1939) that pointed
to the limitations of planning controls for preventing urban sprawl.

Concern mounted and the Committee on Land Utilisation in Rural Areas
(the Scott Committee), reporting in 1942 (Cmd 6378 1942), came to the
conclusion that it should be government policy for agricultural land to have
a prior claim when there was a conflict over land use, so that the country's
potential for food production should not be inadvertently impaired by
short-term economic considerations.

The loss of land, the social price of urban sprawl, and the poor design of
much development all added up to a seemingly powerful case for more
rigorous controls, especially where urban growth threatened good quality
agricultural land. However, there were those who doubted the advisability
of the proposed policies. Denison, in a minority report for the Scott
Committee, emphasised that it was quite possible for agricultural
productivity to outstrip land losses, and that agriculture need not, in
consequence, have a prior right to rural land.

There is considerable support for Denison's view, for although only 80%
of Britain's land surface even then was in agricultural use, it was pointed
out that there was still 'suffecent room and facilities for producing the
amount of food which it is in the social and economic interest of this

country to produce from its own resources' (Wibberley, 1959, p. 77). Certainly the war had shown how the poor production levels on many existing farms could be dramatically improved and many believed that what the farming industry needed was to be more efficient, not to have its land more closely shielded from development.

It was against this background that the Town and Country Planning Act 1947 and other allied legislation were debated, but the government did not fully endorse either side of the argument. As explained in Chapter 5, agriculture and forestry were largely exempted from planning control, but the government did not go as far as accepting the 'onus of proof' argument proposed by the majority of the Scott Committee. Had this been accepted, any intending developer would have had to demonstrate why it is necessary to develop a particular site at a particular time. Instead it has always been government policy to make it encumbent on the planning authority to demonstrate sound planning reasons if it wishes to refuse an application (DOE 1976a). Over the years this has created problems for planning authorities since they often lacked information on which to base a refusal. For instance, it is only very recently that a complete cover of Land Classification maps has been published for England and Wales and even these are at too small a scale to be related accurately to the small sites encompassed by the majority of planning applications. Although MAFF has always been in a position to advise on agricultural land quality, it is only one factor among many and in most cases planning committees appear to have been more strongly influenced by other considerations.

The effect of the planning system on the amount of rural land converted to urban uses can only be measured by surrogate data, the most reliable source being the agricultural returns made by farmers in June of each year for the MAFF. Best has used them to show the extent of change over the past 50 years (Best 1978). The figures must, nonetheless, be treated with some caution, because of differences of interpretation on the part of those submitting the returns and changes in the sample collected over the years. Even so, they reveal sharp enough fluctuations in the rate of conversion to show the underlying changes (Table 6.2). The 5 year mean for the years 1922–26 shows 9100 ha being transferred annually, but this rate then more than doubles to 21 000 ha and remains there until the outbreak of the Second World War. After 1945 the amount of conversion is again much higher than in the early 1920s, but is significantly less than in the 1930s and shows no sign of any sustained upward trend. While this pattern is almost certainly not attributable to any single factor, Best at least is in no doubt about the contribution of the planning system, 'Planning since the war has indeed made a most important contribution towards restricting the spread of urban development onto agricultural land, especially compared with the uncontrolled sprawl of the 1930s' (Best 1978, p. 15). Nevertheless, the annual statistics used to construct the graph in Figure 6.4 show considerable variations from year to year, partly as a result of the data source and partly,

Table 6.2 Annual average net loss of agricultural land to urban use in England and Wales. (Best, R. H. 1978. The extent and growth of urban land. *The Planner* **62**, 15.)

Year	'000 ha
1922–26	9·1
1926–31	21·1
1931–36	25·1
1936–39	25·1
1939–45	5·3
1945–50	17·5
1950–55	15·5
1955–60	14·0
1960–65	15·3
1965–70	16·8
1970–74	15·4

in Best's view, because of the 'stop/go' oscillation of economic controls in the country as a whole. Despite the limitations of the data, Best believes that there is no evidence to support the contention that the threat to farmland from urban development is increasing. Indeed, the area of all rural land lost to forestry each year is substantially more than that lost to building. The rural scene is more likely to be altered by changes in the nature of the agricultural economy, than by extensions to the built up area, as was illustrated in Chapter 4. Nevertheless, the fact that urban uses usually take up lowland farmland makes these incursions far more serious than the moorland losses to forestry, particularly if the potential agricultural productivity of the nation's land is the prime consideration.

Another feature of the Best school is the light it has shed on the regional patterns of urban growth in England and Wales. Until the late 1960s any such analysis was extremely difficult because MAFF change of occupancy data was only published for the country as a whole and not broken down into county sets. The regional figures are now available as 5 year averages from 1945 onwards and Best and Champion have used them to explain the variations in the pattern of urban growth (Best & Champion 1970). They found that the popular conception of a broad belt of urban growth extending from north-west England to the south-east coast was something of a distortion and that a much more accurate picture was of two separate growth areas at either end of the axis. The southerly growth area coincided broadly with the Greater London region and was centred on London itself. In general they found that the rate of conversion declined steadily away from the home counties towards East Anglia, the Midlands, the south-west and the south coast. The northern area, which they called the Central Urban Region, was much more amorphous, comprising several northern and midland conurbations, with the main focus of growth on a line of counties linking the urban areas of Lancashire, Merseyside and the West Midlands. Indeed it was this area rather than Greater London, where urban growth

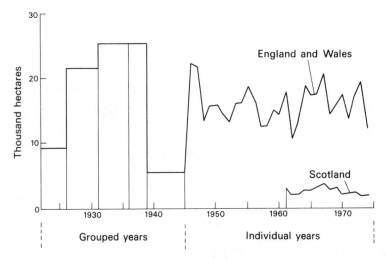

Figure 6.4   Losses of agricultural land to urban use. (Best, R. H. 1977. Agricultural land loss myth or reality? *The Planner* **63**, 15.)

was found to be most rapid. The most interesting part of the study came from a comparison of the amount of urban growth with the figures for regional population increase. Not only were the patterns different, in many respects they were contradictory. Best and Champion attributed the incongruity to the changing standards of space provision in the towns and cities of the Midlands and the north. Before 1945, both population and settlement growth had been rapid in the south, but slow in the north; since then regional policies have evened these growth rates out and planning controls have imposed standard densities of about 30 dwellings per hectare. Furthermore, slum clearance in the north led to the dispersal of people over large new estates in urban fringe areas or New Towns. The study concluded that pre-existing densities have been the single most important explanation of the rate of expansion of urban areas in the post-war period, although other factors include the additional land needed for new factory sites and transport facilities and the combination of natural population growth and immigration.

Finally, Best's and Rogers' work, based on Development Plans and Ordnance Survey maps, has established a clear relationship between the size of a settlement and the space provision per head of population. Although planning policies have succeeded in producing a degree of uniformity in building densities, smaller settlements tend to consume relatively more land for development than larger ones (Best & Rogers 1976). They termed this relationship the density-size rule and postulated that it would hold true generally, even though its applicability was weakening in the specific circumstances of post-war Britain. As far as any verdict on the planning system is concerned, the conclusion is important for it implies that policies

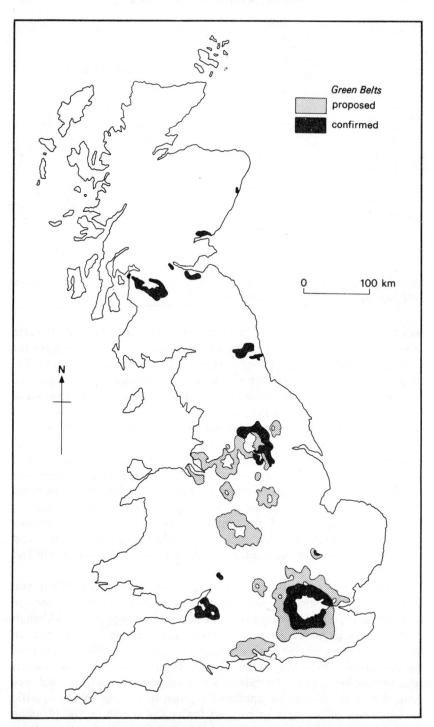

Figure 6.5 Green Belts, proposed and confirmed in Great Britain.

to divert urban growth from cities to rural towns, may in fact have temporarily inflated the quantity of land consumed for urban growth. Subsequently, infill has raised the density in many smaller settlements, so that the discrepancies are narrowing, especially in view of the large areas of derelict land in many city centres.

While there is no doubt about the importance of all this empirical analysis, it was never directly designed to produce a verdict on the efficiency of the planning system for containing urban growth. One study that has

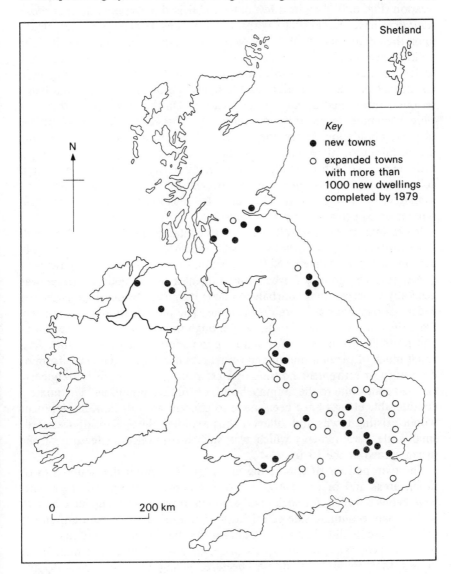

Figure 6.6 New Towns in the United Kingdom.

been specifically concerned with this question and has attempted to interpret the whole urban growth process in England and Wales in terms of the 1947 planning system, is that undertaken by Hall, Gracey, Thomas and Drewett (Hall *et al.* 1974). They argue that the success of the policy of urban containment is evidenced by the generally discrete nature of post-war urban growth. They believe that the growth and coalescence of major towns and cities has been limited by rigidly preventing new development on their fringes, using statutory and non-statutory Green Belts as the main policy weapon (Fig. 6.5). Despite a few much publicised incursions, these *cordons sanitaires* have managed to withstand the considerable pressures placed upon them (Gregory 1970). In fact a policy of urban containment suited the political needs of both town and country remarkably well. The shires were under considerable pressure from their rural electorates to maintain traditional village and small town life styles, which were perceived as being seriously threatened by any urban invasion. On the other hand, the cities were becoming increasingly concerned by the economic impact of the massive population losses they were facing. Rather than encouraging decentralisation, they turned to intensive, high rise development on their expensive urban land, thus sustaining their local tax base and collecting the handsome central government housing subsidies then available, in the process. This also led incidentally, to the physical separation of private and public housing.

Exceptions to the overall pattern of containment were the New and Expanded Towns and various overspill schemes (Fig. 6.6), but they only accounted for a relatively small proportion of new urban development.

Although the policy of urban containment around the urban fringe was generally a success, the suburban character of much new development in the wider countryside was more problematic. The bulk of those moving away from the cities were people who had enough money to be owner-occupiers and could afford the cost of commuting to maintain their urban links. As a result much of the new building in rural areas has either been in the form of dormitory or retirement housing estates, neither of which it is frequently claimed contribute to the ultimate health of rural communities. Fortunately considerable efforts have been made to concentrate this new development within existing towns and villages, thus avoiding the spectre of scattered, small-scale development, which was such an unwelcome feature of the urban sprawl of the 1930s.

The twin policies of containing urban growth within the boundaries of urban areas and of concentrating those decanted into specific locations, were evolved largely for social and aesthetic reasons, but they have had an unfortunate economic side effect in the rapid inflation of land and house prices. Not only did planning policies restrict the amount of land that could be developed, thus forcing up its price, they also limited the number of houses that could be built per hectare, giving a further twist to the inflationary spiral. Nevertheless, housing densities of less than 25 per hectare, common in the interwar years, are now considered lavish and most new residential development, even in villages, is in compact high density estates. Hall and his colleagues believe that the planning system has been successful in terms of the objectives of containment when they say that, 'The policy of urban containment has been applied enthusiastically by the local planning authorities bordering on the main urban areas, particularly the major conurbations, and it has had a clear physical impact. During the 1950s and the beginning of the 1960s, the period between our two land use analyses, these urban areas were increasingly forced to house their surplus populations within their own built up areas' (Hall *et al.* 1973, vol. I, p. 394). Nevertheless, it needs to be remembered that this conveniently ignores the fact that urban areas lost nearly a million people between 1961 and 1971 and that planning is not just a question of urban containment.

Coleman, Director of the Second Land Utilisation Survey, has a rather different interpretation of events. She argues that the planning system was established to improve the quality of housing in England and Wales and, more specifically, to deal with four basic problems: the substandard quality of many existing houses, the lack of amenities such as open space in many residential districts, the sprawling pattern of many residential developments, and the underprovision of dwellings caused by war damage and shortages of building materials. She believes that while there have been improvements, planning has only made a minor contribution. What it has done is to inflate the price of building land, as described by Hall, to breed uncertainty about the future of rural land, resulting in blight and dereliction in much of the urban fringe, and also failed to protect high quality farmland from urban development (Coleman 1976). These conclusions are largely based on resurveys of some of the Second Land Utilisation sheets in the early 1970s. The largest of these areas was the $850 \cdot 5$ km^2 of land in the area of the Thames estuary and the results showed that all new development could have been accommodated on existing waste or scrubland. What happened in fact was that 42 km^2 of farmland were lost and the amount of wasteland virtually doubled to $30 \cdot 5$ km^2. In other words it proved impossible to direct new development to existing areas of unused land, or to prevent further land from becoming derelict. Even Hall has admitted that a high price has been paid, when he concludes that, 'None of this was in the minds of the founding fathers of the planning system. They cared very much for the preservation and conservation of rural England, to be sure, but that was only part of the total package of policies, to be enforced in the interests of all by beneficent central planning. It certainly was not the intention of the founders that people should live cramped lives in homes destined for premature slumdom, far from urban services or jobs; or that city dwellers should live in blank cliffs of flats, far from the ground, without access to play space for their children. Somewhere along the way, a great deal was lost, a system distorted and a great mass of people betrayed' (Hall *et al.* 1973, vol. II, p. 433).

Despite such reservations there is little doubt that were it not for the land use planning controls of the past 30 years, the countryside, especially in the vicinity of the major towns and cities, would be much more urbanised. Both Best and Hall have demonstrated the way in which new building has been held in check. What they have not done is to prove that all is therefore well with the land not built upon. Coleman has shown convincingly that restricting the spread of development is not in itself enough to ensure that farming flourishes in the urban fringe. All too often the result is dereliction, which is less acceptable than the building originally threatened.

Settlement patterns in the countryside

Although key settlement and related policies have been widely adopted in Britain, until recently there has been little significant research into the success or failure of settlement planning (Moseley 1974, Woodruffe 1976). Even the implementing authorities have usually only undertaken the most rudimentary evaluation. For example, in Devon the Second Review of the County Development Plan (Devon County Council 1969), showed that only 17·4% of new development between 1964 and 1968 was located in the 68 key villages, as compared with 22·5% in other villages, and simply blamed this on the effects of outstanding planning permissions predating the plan. No further analysis of the effects of the key settlement policy was attempted, not least because attributing socio-economic trends to specific causes has always been a precarious business and difficulties of interpretation have frequently caused planning policies to be ignored, or even dismissed, in analyses of rural population change. A classic example is Jackson's study of depopulation in the northern part of the Cotswolds (Jackson 1968), which was based on a detailed analysis of census statistics and sought to interpret the demographic changes that had occurred in the inter-censal period 1951–61. As can be seen from Figure 6.7, there was a widespread general decline in population outside the immediate sphere of influence of the four towns at the margins of the study area, Evesham, Stratford-on-Avon, Cheltenham and Oxford. Individually a number of parishes ran counter to the trend, but in each case the anomaly could be explained by special factors. Some of the larger parishes, with populations of more than 2000 in 1951, that were well provided with a service infrastructure had managed to hold their own, especially where the road links to the towns were good, as in the case of the villages to the north of Stratford and Evesham within easy commuting distance of Birmingham. Elsewhere the exceptions could be explained by private employers' 'bus schemes', bringing people from outlying rural areas into work daily in Witney, and by the locations of the military airforce bases. The study concluded that unless employment could be created in such areas the decline would be irreversible and there would be little that planners, or anyone else,

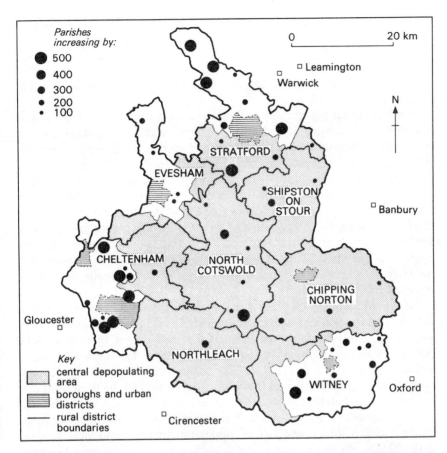

Figure 6.7 Population change in the North Cotswolds 1951−61. (Jackson, V. J. 1968. *Population in the countryside: growth and stagnation in the Cotswolds.* London: Frank Cass.)

could do about it. Indeed, detailed investigation showed that this picture was even more intractable than appeared at first. Over the study period there was considerable substitution of retired people for young people of working age, thus further reducing the potential labour force and making the area still less attractive to prospective employers. Jackson specifically made the point that this movement is difficult to control, however much the planning authority may wish to do so (Jackson 1968). It is also clear that in some respects development control data can be used to distort the true demographic picture. The study revealed a high level of building starts in relation to population growth, but further analysis showed that the trend could be explained relatively easily by lower occupancy levels in the region as a whole, and was not evidence of any incipient economic vitality.

Jackson's work showed how rural settlement change occurred, but it did not really relate this to settlement planning policy. This gap has to some extent been filled by Cloke's examination of the operation of key settlement policies in Warwickshire and Devon (Cloke 1979). Quite fortuitously one of his field study areas partially overlapped the area chosen by Jackson in the Cotswolds, while the other encompassed all four of the areas chosen by the authors for their work in Devon, thus providing a useful basis for comparison.

Cloke's thesis is that the key settlement concept has been applied by local authorities to a number of contrasting rural settlement situations and that it is important to assess the varying degree of success of these policies in the different contexts in which they have operated. He divides the rural areas of England and Wales into two categories, 'pressured' and 'remote', Warwickshire exemplifying the former and Devon the latter. There is little doubt on the basis of other evidence (see Ch. 4) that this division is far too crude and simplistic, representing only the two extremes of the rural situation and none of the more subtle variations in between, but the contrast is nonetheless important. It needs to be pointed out however that only parts of the north and west of the County of Devon qualify as areas of remote rural population.

In Warwickshire a two-tier version of a key settlement policy was inaugurated in 1966 and became fully operational in 1971. Twelve villages were identified as suitable for 'moderate expansion' of more than 1000 people by 1986, while a further seven were to be allowed 'modest expansion', in the range 100 to 500 persons (Fig. 6.8). The proposed concentration of development in a county dominated by the West Midlands conurbation is therefore very considerable, especially in view of the demand for rural housing in any area within commuting distance of one of the major cities. Cloke's analysis of population change and new housing development reveals that the policy so far has met with only mixed success. Undoubtedly the policy cuts across the natural pattern of demand and many villages, which clearly have the capacity for growth, are being held back due to the fact that they are not key settlements. On the other hand the existence of the policy has not succeeded in confining growth within the key settlements entirely, even in the West Midlands Green Belt where there is a strong presumption against development. In the south of the county, along the Cotswold edge, there is the reverse problem with designated villages failing to achieve their growth targets.

The distribution of new house-building is somewhat more related to the policy, but there are still important anomalies in both the public and the private housing sectors. For example, a considerable amount of new private housing has been built in non-selected villages in the area around Warwick and Leamington Spa and in the buffer zone between Birmingham and Coventry. Council houses conform more closely to the guidelines, but a number of houses were still built in non-key settlements and six key settlements have received no new council houses at all. Three of these, Kineton, Bishops Itchington and Stratton-on-Dunsmore, were in the south

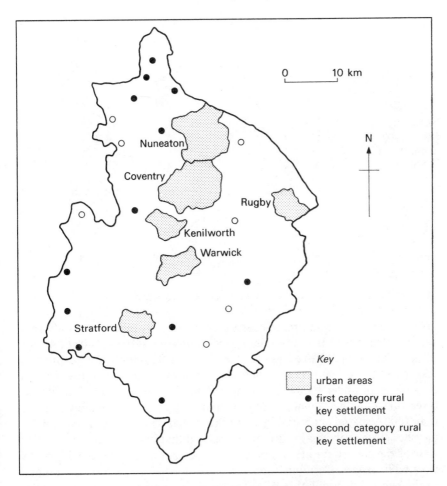

Figure 6.8 Key settlements in Warwickshire. (Cloke, P. 1977. *The use of key settlement policies in the planning of rural areas.* Unpub. PhD thesis, University of London.)

of the county and were among the villages cited above as having not so far reached their growth targets.

Any final conclusions about the Warwickshire key settlement policy are premature since it has been designed to operate up to 1986. Nevertheless, it appears to have only had limited success in diverting growth to the chosen locations so far and, even more important, does not appear as yet to have penetrated very deeply into the consciousness of either developers or the public at large.

In Devon a key settlement policy has been operating for much longer. It was first proposed in the First Review of the County Development Plan in 1964, when 68 villages were designated, the number being reduced to 65 at the time of the Second Review in 1970. The total number of settlements is

Table 6.3 Population change in key settlement parishes 1961–77 (by number of parishes). (Cloke, P. 1977. *The use of key settlement policies in the planning of rural areas.* Unpublished thesis, University of London.)

Population size	Total	under −25	−24 to 25	26 to 100	101 to 500
101–500	12	2	7	3	
501–1000	25	3	6	11	4
1001–5000	32	8	5	4	13
more than 5000	1				
Total	70	13	18	18	17

	501 to 1000	more than 1000
101–500		
501–1000	1	
1001–5000	2	
more than 5000		1
Total	3	1

also very much greater than in Warwickshire and, since they are also evenly distributed throughout the county (for reasons explained earlier, see p. 120), in the remoter and more sparsely populated areas some are only very small settlements. Thirty-seven had populations of less than 1000 at the time of designation and twelve less than 500. Up to 1971 population growth was disappointing as shown in Table 6.3 and only five of those with a population of under 1000 grew by more than 100 persons, and of the thirty-three villages with populations of more than 1000 only sixteen grew by more than 100. A spatial analysis of growth shows that it mostly occurred in those parishes close to the major urban centres and that in the remoter north and west of the county the general pattern of decline was only reversed in a few isolated instances. Certainly there is no sustained evidence to suggest a halt to the general depopulation of the more isolated areas.

Not surprisingly the pattern of new house-building is broadly similar, key settlements in the south and east of the county attracting much more new development, particularly new council housing, than other areas. As in Warwickshire, however, the main problem is that in the less attractive areas many of the selected villages are too small to act as foci for growth, even with the stimulus of being designated a key settlement. At the other end of the scale, again as in Warwickshire, there are a number of villages within commuting distance of Exeter, Plymouth and Torbay, where the demand for expansion is being frustrated through rigid adherence to the policy.

In conclusion Cloke's work suggests that settlement policies have at most had a marginal effect on both population and housing, particularly in halting decline in the remoter rural areas, although in the pressured areas they have been more successful in holding back growth in some of the undesignated settlements where there was a demand for expansion. Indeed, both he and Jackson agree that the most successful settlements are the large ones that are already thriving and that, in most cases, they would have continued to grow and would have been selected for development in any case. This is a conclusion also borne out by a study of Ivybridge, a settlement within commuting distance of Plymouth that has experienced explosive growth since the early 1960s (Jones 1977).

One of the limitations of Jackson's and Cloke's work, based on parish statistics, is the lack of correspondence with the actual built-up area of settlements. Many rural parishes are very large and can give a rather misleading picture of the pattern of development. One way of overcoming this difficulty is by analysing the distribution of new building from planning applications, as the authors have done in the four areas in Devon referred to in Chapter 5. Although the technique is laborious and time-consuming, it does have the advantage of allowing individual development proposals to be more precisely located and, by analysing the pattern of refusals and permissions, of enabling an assessment to be made of the way in which planning authorities have been able to modify the pattern of demand. By giving each application a National Grid reference number it is easy to manipulate the data into map form based on kilometre squares and Figure 6.9 shows the distribution of outline permissions in east Devon, Dartmoor, Exmoor and mid-Devon for the period 1964–74. It also identifies those squares in which there are recognisable settlements and differentiates these into key and non-key settlements. The village areas are not precise delineations of land that has been built upon, but rather indications of those grid squares containing all or part of a settlement.

The most striking feature of the east Devon distribution is the very large number of villages, reflecting long-standing rural and suburban pressures for development in this area. Only on the string of commons, occupying the north–south running ridge across the middle of the area, is there any real reduction in the amount of development. Since there are so many villages, it is hardly surprising that 81·3% of outline planning permissions have been located within one or other of them. On the other hand the evidence for a particular concentration on the two key settlements is less clear. Woodbury has undoubtedly acted as an important focus in the west of the area, attracting 18·7% of all development, but in the east Newton Poppleford has not established anything like the same domination (Blacksell & Gilg 1977). The reasons for this discrepancy are not clear, but the situation underlines the fact that designation by no means automatically ensures that a settlement will become a local growth pole. Nevertheless, over the area as a whole, the small quantity of new building that has taken place in the open countryside (18·7%) is encouraging, especially in view of the fact that much of it is accommodation for farm workers, which the planning authority is virtually powerless to refuse even if the validity of many of the claims is subject to doubt.

On Dartmoor there are fewer individual villages but three key settlements, Moretonhampstead, Dunsford and Christow. Despite this

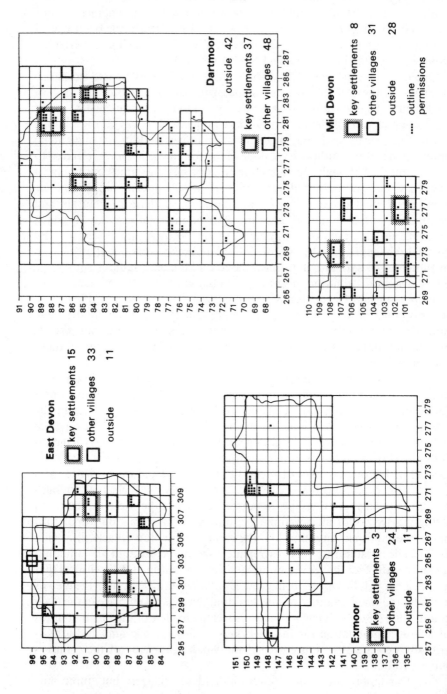

Figure 6.9 Unweighted outline planning permissions in the four study areas 1964–74. (Authors' survey.)

small number, the key settlements have not been notably successful in dictating the distribution of new development, only attracting 22·0% of outline permissions, as compared with 33·9% in the non-key villages. In fact, with the exception of Moretonhampstead, which is nearly three times as large as any other village, neither of the other two key settlements was an obvious choice. Ilsington, Lustleigh or Bridford could have laid just as strong a claim and the fact that they were not selected in no way diminished the demand for housing there.

The other striking feature about development on Dartmoor is the very large amount of new housing that has been permitted in the open countryside. Of all permissions 44·1% were not within the confines of any village, a higher proportion than any of the other areas studied. The explanation for this apparent anomaly is the very heavy demand for housing from commuters living in Exeter and Torbay and the needs of agriculture. Farms in this part of the National Park are very scattered and short of modern residential accommodation, making it necessary to build substantial numbers of new houses. Unfortunately, the situation has produced a clear conflict between the aims of the settlement policy and the needs of the agricultural industry.

Exmoor presents a complete contrast to the above two areas, with much less demand for development and a distinctly urban character to its largest settlement, Lynton-Lynmouth. Although little more than an overgrown village, the nineteenth-century holiday resort enjoyed the status of an Urban District prior to local government reorganisation in 1974 and was, therefore, not eligible for consideration as a key settlement but it was instead designated as a 'Coastal Resort'. This has helped it attract the bulk of new development and the contrast between its success and the stagnation in the one key settlement, Parracombe, is most striking. Parracombe is one of those villages selected more for political, than sound socio-economic reasons and its failure bolsters the argument that key settlements tend to confirm existing trends, rather than initiating new ones. A striking feature of the distribution of permissions on Exmoor is the relatively small proportion, 28·9%, in the open country. Since it is an area with few villages and many isolated farms, the extent to which development has been confined is surprising and in marked contrast to Dartmoor.

Table 6.4 An analysis of dwellings in the south-east Dartmoor, mid-Devon and western Exmoor study areas, by age. (Authors' survey.)

Date of construction	*South-east Dartmoor*		*Mid-Devon*		*Western Exmoor*	
	no.	*%*	*no.*	*%*	*no.*	*%*
pre-1914	1406	56	771	53	743	73
interwar	437	17	160	11	101	10
1945–63	418	18	248	17	121	12
1964–74	272	11	268	17	57	5

In all the three areas described so far the need to protect the aesthetic qualities of the landscape is a major consideration in determining development control issues, but in mid-Devon there are no such formal constraints. Although it is an attractive area with a number of picturesque villages, none of it is covered by any kind of landscape protection order and the emphasis is strongly on the protection of agriculture and related industries. There are two key settlements, Lapford and Copplestone, which between them have attracted 8·1% of the outline permissions, as compared with 50·8% in the non-key villages and 41·0% in the open countryside. In other words there is little evidence of the settlement policy having much direct effect on the location of new housing.

When considering the effects of planning decisions, there is always a danger of getting things out of proportion. Despite all the new development that has occurred since the end of the Second World War, the majority of houses in rural areas date from before the First World War and the built environment is a long way from being dominated by modern designs and materials as shown in Table 6.4. Nonetheless, just because of this predominance of old buildings, it is important that new development merges with them sympathetically and policies must be flexible enough to account for the peculiarities of each individual area. Much of the building that has apparently been permitted in the open countryside, particularly in Dartmoor and mid-Devon, can be justified as the infilling and rounding-off

Table 6.5 Ratio of refusals to permissions of outline planning applications in the east Devon, south-east Dartmoor, western Exmoor and mid-Devon study areas. (Authors' survey.)

	Permitted	*Refused*
East Devon		
total	1	3·9
key settlements	1	2·7
non-key settlements	1	2·5
elsewhere	1	7·1
South-east Dartmoor		
total	1	2·8
key settlements	1	1·6
non-key settlements	1	1·7
elsewhere	1	6·9
Western Exmoor		
total	1	1·2
key settlements	1	1·3
non-key settlements	1	1·2
elsewhere	1	1·4
Mid-Devon		
total	1	0·6
key settlements	1	0·5
non-key settlements	1	0·7
elsewhere	1	0·5

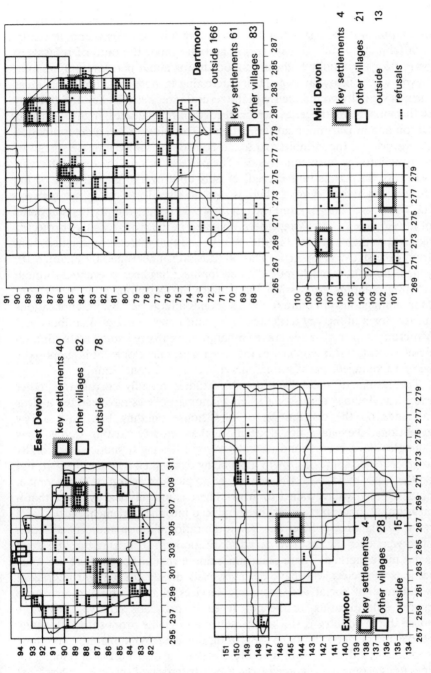

Figure 6.10 Unweighted planning refusals in the four study areas 1964–74. (Authors' survey.)

of developments that were only partially completed when the settlement policy was initiated.

Furthermore, the planning authority ought to be judged not only by the development it has permitted, but also by what it has refused and, as can be seen from Table 6.5, in east Devon and Dartmoor the ratio of refusals to permissions was quite high, in Exmoor it was about par, and only in mid-Devon was it below average. It is interesting to note that in all four areas there is virtually no difference in the refusal rates between key and non-key settlements, but in the 'elsewhere' category the refusal rate is higher – in east Devon and in Dartmoor substantially so. This must reflect a determination on the part of the authority to try to control the indiscriminate spread of housing into the open countryside. The pattern for mid-Devon is different. Not only do permissions exceed refusals both overall and in each of the three individual categories, there is also no significant difference between the 'elsewhere' category and the rest. It would appear that tighter standards of development control have been operated in the more sensitive protected landscapes than elsewhere in the county, and that these have shielded them from excessive development. This conclusion is underlined by Figure 6.10, which shows how much sporadic development has been prevented through refusing planning permission, particularly in Dartmoor and east Devon. More ambiguously, Figure 6.10 also reveals a large number of applications turned down in the key settlements. In some cases, notably Woodbury and Moretonhampstead, a sewerage embargo has enforced some restriction on development, but it emphasises that even where development is supposedly being encouraged, permission is never a foregone conclusion.

The results of these four detailed studies broadly corroborate other empirical findings discussed earlier: planning policies have exerted a clear influence on the distribution of new house building but, with a few exceptions, they have not done more than modify existing trends. Few examples have been found of villages experiencing a sudden and totally unprecedented spurt of growth after being designated a key settlement; but on the other hand, with the one rather atypical exception of Parracombe, no designated village failed to grow. There is unequivocal evidence that a measure of concentration has been induced into the settlement pattern as a result of planning policies but, given the rationalisation and centralisation of services over the past two decades, it is perhaps surprising that the trend is not more marked. There seems little doubt that although there are some sections of society who have been seriously disadvantaged by the gradual withdrawal of social and economic services in rural areas, others have proved able to adjust to the new levels of provision relatively easily. A house in the country is still an attractive and viable proposition for many people and demand shows no signs of diminishing. What planning has done is to hold that demand within bounds, so that the distribution of population does not become too divorced from the distribution of services. In National Parks and other aesthetically valued landscapes it has also attempted, with

Table 6.6 The decline of rural services in the south-west 1972–77. (Standing Conference of Rural Community Councils 1978. *The decline of rural services*. London: the Standing Conference.)

Percentage of villages with each facility in 1972 and 1977

	Primary school		Post office		Shop		Chemist		Doctor's surgery	
	1972	1977	1972	1977	1972	1977	1972	1977	1972	1977
Avon*	81*	78	78	74	75	71	52	50	—	—
Gloucestershire	76	74	91	85	82	71	14	14	33	32
Wiltshire	67	61	84	78	78	68	5	4	24	23
Somerset	62	61	87	85	84	79	8	8	33	32
Cornwall	(1967) 61	(1976) 56	(1967) 83	(1975) 75	—	—	—	—	(1967) 32	(1976) 27
Devon	—	—	73	72	82	72	—	—	(1967) 27	(1975) 24

Percentage decline for each facility

	Period	School	Post office	Shop	Chemist	Doctor's surgery
Avon	1972–77	− 4·0	− 4·5	− 6·0	− 4·5	—
Gloucestershire	1972–77	− 2·5	− 8·0	−13·0	nil	− 3·0
Wiltshire	1972–77	− 9·0	− 8·0	−13·0	−14·0†	− 2·0
Somerset	1972–77	− 1·0	− 2·5	− 5·5	nil	− 5·0
Devon	1967–75	—	− 1·5	−11·0	—	−14·0
Cornwall	1967–76	− 7·5	−10·0	—	—	−19·0

* 'Primary' not specified in Avon survey.
† Only one closure.
— Indicates that figures are not available.

some considerable success, to ensure that new development does not intrude into the established scene.

To consider the effectiveness of key, or for that matter any other, settlement policies purely in terms of new housing is to do them less than justice. The influence that a planning authority is able to exert through the development control system is strictly limited to reacting to the initiatives of others, but by publishing the policies the planners also hope to persuade the providers of services to concentrate their investment in the selected settlements in the expectation that householders will gradually follow suit. The problem is that while there is ample evidence for the rationalisation of services, the reaction of residents is more equivocal. Houses cannot simply be moved from one place to another and new development has to be sited where there is land available, so that there is always pressure for services to match the distribution of housing rather than the other way around.

Whatever the rights and wrongs of that particular debate, there is no doubt that the overall level of rural service provision has declined sharply. Greater mobility for the majority of the population has reduced the demand for shops and other services in isolated locations, and economies of scale mean that services are more efficiently provided from large centres. As a result many village police, fire and ambulance stations have disappeared, along with other public and private services. Table 6.6 illustrates the trends and shows that all have declined although the state of decline varies both between services and between counties. The survey, which was conducted by the Standing Conference of Rural Community Councils, shows that shops have been badly affected in all areas and that the numbers of doctors' surgeries have fallen sharply in Devon and Cornwall, though not elsewhere. The county that has been hit worst overall is Wiltshire, suffering high rates of loss for primary schools, post offices and shops.

The true meaning of such losses only becomes apparent when related to actual services in real villages. The report cites the example of Upottery, a key settlement in east Devon but outside the authors' study area (Standing Conference 1978). When designated in 1964, it boasted a public house, a church, a village hall, a primary school, a post office-cum-shop-cum-garage and a daily bus service to the nearby market town of Honiton. The post office and its attendant facilities have subsequently closed and the bus service has dwindled to one day a week. Now, if passengers wish to make the return journey on the same day, they are able to spend just one and three-quarter hours in Honiton. The only compensation for Upottery is that another sub-post office has opened, but it is three-quarters of a mile from the centre of the village and has none of the extra retail facilities provided by the former establishment.

While such a picture is not unusual, Cloke has shown that in both Warwickshire and Devon the services in the key settlements have in general either held their own, or even marginally increased (Cloke 1979). He also shows, however, that elsewhere there has been an almost universal decline.

At first sight this may appear to vindicate the policies, but such a conclusion does not bear detailed analysis. In both counties there has been an overall reduction in the level of public and private services and the key settlements have done little more than hold their own. There is almost no evidence for the concentration of investment that was supposed to occur, and it is doubtful whether people living in key settlements have derived much benefit from the new order, they have just suffered less of a decline.

The problems of the declining areas and the extent to which the key settlement policy has managed to help them, were at the heart of a study of the remote mid-Devon village of Hatherleigh, commissioned by Devon County Council from Exeter University (Jones 1979). It sought to show what effect the social and economic changes in the 1960s and 1970s had actually had on such an isolated village community, particularly in view of its key settlement status. The study showed that in the opinion of the residents two problems were of overriding importance: the need for a by-pass, so that the mediaeval fabric of the village would no longer be forced to suffer the pressures of twentieth-century traffic; and the creeping isolation of the population from the urban services on which they were, ironically, becoming ever more dependent. The popular view was that official planning policies had done nothing to solve either, they had simply raised expectations and then failed to produce results. A by-pass was written into both the First and Second Reviews of the County Development Plan, but no firm date has yet been given for the commencement of the new road, despite the fact that the existing route through Hatherleigh was officially upgraded in 1977 by the county council to the status of 'primary standard route' on the functional route network (Devon County Council 1979). The alternative has been to adopt temporary piecemeal solutions in the village itself so that the greatly increased volume of heavy goods traffic between north and south Devon may pass through. As a result buildings on the main street have become structurally dangerous and have had to be demolished, leading directly to the disappearance of several shops, including the only chemist in Hatherleigh. To blame these closures solely on the failure to build a new by-pass is of course to oversimplify, but there is little doubt that the general deterioration in the village's environment caused by the continuing traffic problem has impaired its function as a rural shopping centre.

Other services, notably the provision of medical facilities, have also declined sharply in Hatherleigh over the past decade. Historically the nearest hospital was in Okehampton 7 miles away and, as there was a frequent bus service, all the published plans for the village envisaged this satisfactory arrangement continuing. The Regional Health Authority, which is independent of the County Council, has decided to concentrate care in large district hospitals and there are only four in the whole of Devon, in the main urban centres of Barnstaple, Exeter, Plymouth and Torbay. Although the hospital in Okehampton has not actually closed, it no longer has surgical facilities and the Regional Health Authority is threatening to withdraw the maternity unit. Patients requiring anything other than routine treatment are now sent either to Exeter, Barnstaple or Torbay, all of which are about 30 miles from Hatherleigh, and difficult to reach by public transport.

Similar rationalisation has affected other services, such as ambulances and the police and the power of local authority planners to influence the pattern of events is strictly limited. They do not control many essential services and regional authorities, like those for Water, Health and the Police, are subject to different pressures and may not accept local authority planning criteria. Even within the local authority itself it is by no means easy to co-ordinate policies, particularly when it comes to timing new initiatives. Housing, schools and other developments apparently under their direct control are also heavily dependent on outside agencies. For instance, the expectation that all modern houses have mains water and are connected to a water-borne sewage system often gives the Regional Water Authority a power of virtual veto over new development. In many rural areas, and Hatherleigh despite its key settlement status is a case in point, facilities are not available and, even where they have been provided, they are frequently already being used to capacity. Gilg has shown elsewhere (Gilg 1978a) that inadequate sewage facilities imposed an embargo on development in 34 of the 68 key settlements in Devon, and a similar situation has been reported in other rural areas.

The Regional Water Authorities are naturally very much aware of the problems that they have inadvertently caused, but can only remedy the situation slowly within the limits of available funds. The Structure Plan actually cites where the absence of sewerage and other facilities is going to hold up implementation (Devon County Council 1979). In the meanwhile the availability of excess sewage and water supply capacity in villages not selected by the planning authority for further development has led to great pressure for planning permission, which has been hard to resist. The authorities have found themselves in the unenviable position of not being able to sanction new building where they had deemed it should be located, and virtually having to grant permission where they had deemed it should be discouraged.

It is easy to unearth the kinds of hiatus and incongruity described in this section, but realistically strategic policies, whether for key settlements or anything else, can never, nor should ever be anything other than a potential blueprint for development. Deviations were to be expected on many fronts, just as has happened, and at least the existence of the policies allowed them to be recognised as such. Unfortunately, once a policy has been published, it is expected to be implemented in full instantly, yet patently this cannot happen. The change is bound to be slow and evolutionary, particularly where planning is concerned, as it may take several years for existing permissions to be translated into buildings on site. It also takes a very long time for any plan to be fully understood by those trying to implement it and the pressure for instant success has sometimes meant that policies have been abandoned before they have had a chance to succeed. Nevertheless, too much was expected from the planning system in rural areas. Land use considerations are only one element in a planning strategy; new developments are influenced by many other criteria as well and this is especially true of services, like hospitals, water supply and sewage disposal, where the authorities are under great financial pressure to plan for the maximum return on their investments irrespective of planning policies.

Village layout and building design

Central govenment circulars, advising local authorities on how to implement planning powers, have since the 1950s placed increasing emphasis on using the development control system for modifying the detailed location, design and siting of buildings. A great deal of time is spent at both officer and committee level in discussing modifications to planning applications, so that they do not intrude on the surrounding environment. In rural areas particular emphasis has been put on this aspect of the planning system. One of the express purposes of National Parks, Areas of Outstanding Natural Beauty and other similar landscape designations is to ensure that even higher standards of development control are adopted in such areas than elsewhere. In general these exhortations are not accompanied by any extra powers or statutory procedures, except for the preservation of historic parts of the built environment. Under the terms of the Civic Amenities Act 1967, local authorities can designate valuable and sensitive sections of settlements as conservation areas. This means that not only is new development strictly controlled, but existing buildings can only be altered, even by painting, with the consent of the planning authority. There is also the power to list buildings, which again means that modifications to the structure require the consent of the planning authority. These conservation powers are very important because they not only enable the planning authority to vet new development proposals, they also provide money for initiating improvement schemes in the areas in question.

Planning authorities in both urban and rural areas have made good use of these powers and conservation areas of all shapes and sizes are now a common feature of cities, towns and villages throughout England and Wales. As can be seen from Figure 6.11, extensive use has been made of these powers in Devon where there are 152 conservation areas, 19 of them recognised for the purposes of grant aid by the Department of the Environment, and 6 town schemes, one of which, Totnes, has set standards for the whole of the country.

Preservation of an existing group of buildings is relatively easy in that there is little dispute about the ultimate objective. When it comes to new building no such consensus exists and planners have been reluctant to

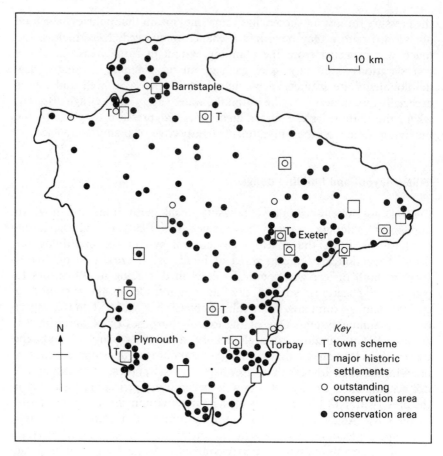

Figure 6.11   Urban conservation policies and proposals in Devon. (Devon County Council 1979. *The County Structure Plan.* Exeter: The Council.)

commit themselves on standards, especially in rural areas. Thorburn, whose guide to village planning has already been referred to in Chapter 5, warns that, 'It is well to be wary of the cure-all prescriptions for better village design, as often these are no more than personal opinions. All too frequently they are based on a particular ideal, appropriate in one or two villages, or even in a whole district, but quite unjustified elsewhere' (Thorburn 1971, p. 84). He himself then partially falls foul of his own strictures in that he attempts to set out a series of design objectives, covering such things as materials, building form, layout, tree planting, density and circulation. In the main, however, planners admit that design policies for rural areas have relied heavily on concepts like traffic and pedestrian separation and comprehensive estate development, which were evolved to cope with urban and suburban rather than rural environments. One notable problem has been building regulations, which often impose for safety reasons urban solutions on small-scale village layouts. Much official government guidance has also relied on the use of building densities. Gilg has elsewhere described density as a sterile concept and it is manifestly true that good design is not limited to any particular number of houses to the hectare (Gilg 1978a). There are many examples of good, bad and indifferent design in the countryside at high, low and medium densities.

It is difficult to summarise the design faults of new building in rural areas, but Green has identified three fundamental areas for concern. He argues that, 'detailed layouts have failed to take advantage of the overall pattern which has been laid down because:

(a) by far the greater amount of new development is dull and monotonous, and layout and individual house design is singularly lacking in inspiration;
(b) many layouts approved by local planning authorities lack either open space or provision for planting trees;
(c) the lack of footpaths, linking culs-de-sac, loop roads, etc. means that hundreds of yards are added to each walk to and from schools, shops and bus stops' (Green 1971, p. 72).

A classic example of the unsympathetic way in which some villages have

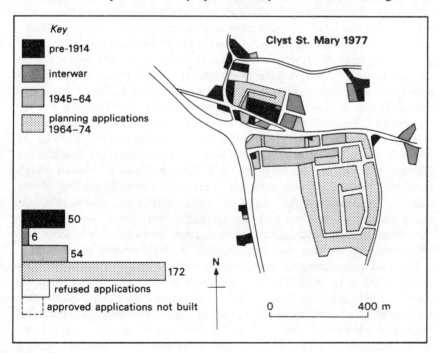

Figure 6.12 Development in Clyst St Mary, Devon, actual and proposed. (Authors' survey.)

been allowed to grow is provided by Clyst St Mary on the outskirts of Exeter. The pre-1914 village contained only 50 houses and huddled close to the narrow road between Exeter and Sidmouth as shown in Figure 6.12. The village hard.y changed in the interwar years with only six further dwellings being added, but in the 1950s a much needed by-pass was finally completed. Since then a succession of large planning permissions have been granted and 226 new dwellings have been added, the majority of them on the far side of the by-pass from the original village, thus creating a new appendage out of all proportion to the initial development and separated from it and its services, including a primary school, by a busy main road. An analysis of the planning applications for the village shows that very few were turned down, so that the authority does not appear to have done much to resist the development pressure in Clyst St Mary, despite the fact that it was not designated a key settlement and is situated on Grade 1 and 2 land. What happened is by no means typical of village development in Devon or elsewhere, but it does illustrate how easily a situation can get out of hand. Not only has Clyst St Mary grown nearly five times in size, the new development is isolated from the original village by the by-pass, yet still relies on many of its services, such as the primary school, that are found there. The lack of co-ordination between users and the providers of services and the way in which the recent building has swamped the original settlement raise serious doubts as to whether the planning system played any constructive role in the development whatsoever.

In an attempt to overcome some of these problems, both central and local government have published a whole series of building guides. Nationally, as already mentioned, these have leant heavily on the concept of building density. Locally, the quality and the nature of the advice has varied from one authority to another, but some guides, such as the *Design guide for residential areas*, published by Essex County Council in 1972, have had an influence well beyond the area for which they were conceived. Most of the National Parks have had building guides for many years, the first one for Dartmoor appearing as early as 1955 (Devon County Council 1955). However, the tendency in these has been to err heavily on the side of caution, emphasising the need to make any new development as inconspicuous as possible and urging that it conform with a particular traditional building style, even though this may be based on a fanciful pre-conception, rather than the actual facts of the built environment.

One area where there has been specific and imaginative guidance on new development is farm buildings. There are a number of excellent manuals published by various government departments illustrating how new building techniques need not necessarily produce ugly intrusions into the rural landscape, especially if the location and the use of materials are considered sympathetically (MAFF 1972b). Unfortunately the very limited planning controls over agricultural building mean that this excellent advice is all too often ignored.

Figure 6.13 New residential development in the Exmoor and Dartmoor National Parks, illustrating the lack of imaginative design. (Authors' photographs.)

Evaluating the quality and suitability of designs for new buildings is extremely difficult. As in all matters of taste 'one man's meat is another's poison'. All too often it is impossible to distinguish between new development in National Parks and the most stereotyped suburbia. The selection of photographs shown in Figure 6.13 taken in the Exmoor and Dartmoor National Parks show new residential development which, from a design point of view, could have been located anywhere. The layout and the materials used are totally unremarkable and there is no evidence of the higher standards that might have been expected in such an area.

To investigate the problem in more detail the authors attempted to relate the design of new development to that of the existing building stock in the Exmoor and Dartmoor National Parks and in mid-Devon. The aim was to try to quantify the extent to which building styles and materials had changed since the planning system was inaugurated and to try to establish whether there had been more success in adhering to traditional building practices in the National Parks.

A survey was made of all the houses in the three areas and data were collected on the date of construction, the type of dwelling (terrace, semi-detached, detached), the number of storeys (bungalow, 2 storey, 3 or more storeys), building materials (brick, stone, render, cob, other), and roofing materials (tile, slate, tin, thatch, asbestos, wood, tarpaulin, mixture), as well as a number of other characteristics that have not been included in the present analysis. The first fact to emerge is the extent to which the nature of the present built environment in all three areas was determined before the First World War. As can be seen from Table 6.7, in Exmoor 73% of houses date from this period and in Dartmoor and mid-Devon the proportions are 56% and 53% respectively. One is looking at an historic landscape that has to be adapted to changing social and economic circumstances.

These results strongly contrast with the 1977 National Dwelling and Housing Survey, that found that in the country as a whole only a quarter of all dwellings were built before 1919 and that a minority of these are terraced houses. Of the rest, a quarter were built between 1919 and 1939, the majority being semi-detached houses and the remining half dated from the years since the last war. As the following analysis shows, this suggests a different evolution of building design in rural areas, both before and after planning controls were introduced.

Even in the three study areas in Devon, however, at least 15% of houses have been built since the planning system in its present form came into existence and in mid-Devon the proportion is as high as 36%. Thus the post-war developments have had a substantial impact here as well although they only account for a minority of houses, and in the two National Parks the relatively small numbers of new houses means that the existing landscape has, by and large, been preserved.

Controls have not prevented marked changes in the design of new buildings. Table 6.7 shows that in all three areas, two or more storey

Table 6.7 Building stock in the south-east Dartmoor, mid-Devon and western Exmoor study areas, classified according to date and mode of construction (%). (Authors' survey.)

(1) Date of construction

	Dartmoor	Mid-Devon	Exmoor
pre-1914	56	53	73
interwar	17	11	10
1945–63	18	17	12
1964–74	11	19	5
total	100	100	100

(2) Mode of construction (i)

	Dartmoor				Mid-Devon				Exmoor			
	terrace	semi	detached	total	terrace	semi	detached	total	terrace	semi	detached	total
pre-1914	41	12	47	100	41	18	41	100	46	11	43	100
interwar	15	14	71	100	16	24	60	100	12	14	74	100
1945–63	49	20	31	100	49	25	26	100	10	51	39	100
1964–74	6	19	75	100	19	20	61	100	9	4	87	100

(3) Mode of construction (ii)

	Dartmoor				Mid-Devon				Exmoor			
	bunga-low	2 storeys	3+ storeys	total	bunga-low	2 storeys	3+ storeys	total	bunga-low	2 storeys	3+ storeys	total
pre-1914	4	92	4	100	1	97	2	100	4	79	17	100
interwar	40	59	1	100	32	68	0	100	30	68	2	100
1945–63	32	68	0	100	33	67	0	100	52	46	2	100
1964–74	79	21	0	100	68	32	0	100	49	51	0	100

buildings were traditionally the norm, but they have now been superseded by the bungalow, a phenomenon that was largely unknown prior to the First World War. Terraced housing, which appears to have been almost as much a feature of rural as of urban Britain in the nineteenth century, has been replaced by detached and semi-detached dwellings, especially in the period since 1964. In other words the density of development appears to have been reduced markedly. It should be remembered, however, from the analysis of the distribution of building referred to earlier in this chapter, that there has not necessarily been any great dispersal of housing in the counryside. In general there was ample land on the fringes of and within existing settlements to accommodate the new and more space-consuming forms of house building.

The analysis of building and roofing materials shown in Tables 6.8 and 6.9 is difficult to interpret, in that there were important differences between each of the three areas prior to the First World War, reflecting the local availability of certain materials. In both Dartmoor and Exmoor a higher proportion of buildings have traditionally been stone built than in mid-Devon and, while slate predominates as a roofing material everywhere, there is noticeably more thatch in mid-Devon and Dartmoor than on

Table 6.8 Building materials for residential development in the south-east Dartmoor, mid-Devon and western Exmoor study areas. (Authors' survey.)

Materials in %	South-east Dartmoor		Mid-Devon		Western Exmoor	
	pre-1945	post-1945	pre-1945	post-1945	pre-1945	post-1945
brick	4·3	4·5	5·8	9·7	1·0	9·0
stone	25·6	5·3	12·4	2·0	41·6	2·8
render	67·4	85·6	80·1	71·5	47·7	73·8
cob	0·1	—	0·4	—	—	—
other	37·5	4·6	1·2	14·9	9·8	14·5
	100·0	100·0	100·0	100·0	100·0	100·0

Table 6.9 Roofing materials for residential development in the south-east Dartmoor, mid-Devon and western Exmoor study areas. (Authors' survey.)

Materials in %	South-east Dartmoor		Mid-Devon		Western Exmoor	
	pre-1945	post-1945	pre-1945	post-1945	pre-1945	post-1945
tile	21·3	94·0	15·5	93·9	18·1	79·8
slate	62·8	3·5	49·8	3·3	79·4	6·7
tin	1·2	0·6	1·1	0·6	0·6	0·5
thatch	13·1	—	27·1	0·2	0·5	—
asbestos	0·3	1·0	3·7	0·4	0·7	1·0
wood	—	0·1	—	1·5	0·8	3·5
tarpaulin	0·1	—	—	—	0·1	—
mixture	1·2	0·7	2·7	3·0	—	8·1
	100·0	100·0	100·0	100·0	100·0	100·0

Exmoor where it is virtually non-existent. The reason for this particular discrepancy is that slate is quarried extensively in western Exmoor, but there is none in mid-Devon and little in eastern Dartmoor.

As with the type of building and the mode of construction, the era of planning has brought with it important changes to both building and roofing materials. Render has everywhere tended to replace stone walls and tiles both thatch and slate. It would appear that planning could do little to overcome the relative economic advantage of more modern materials, even in National Parks where conservation subsidies are available. Nevertheless, there is some evidence that temporary materials, such as wood and asbestos, have been reduced in the last 30 years, especially in Dartmoor. This is an achievement, for it reflects the higher building standards that the planning process has managed to encourage.

The results of the survey bear out, therefore, the results of the photographic evidence referred to earlier in this section. Unquestionably there has been a vast improvement in the technical quality of building in rural as in urban areas since effective planning and building regulations came into force, but there is little evidence that the planning system has prevented the general introduction of urban house styles into the countryside. Green's reservations about the appropriateness of much building for rural areas seem well founded: traditional house forms and designs are not generally being perpetuated in new development, even in areas where there is a premium placed on conservation. However, Tables 6.8 and 6.9 do show that a traditional building design may be hard to define and the myth of the ubiquitous thatched granite Dartmoor cottage is lain bare by these figures.

Large-scale developments

Although most planning applications are for individual new buildings or modifications to existing ones, there are a few exceptions. From time to time, particularly in rural areas where there are large tracts of land that have not been built upon, an application is made for a development of such large-scale that it is likely on its own to significantly alter the landscape. Most, such as new reservoirs, arise from settlement growth elsewhere and are intended to serve the new demands.

Technically size has no bearing on the way in which projects such as reservoirs, major industrial developments and extensive recreation schemes are considered by the planning system, but in practice the decision-making process in such cases tends to be quite different. Invariably there is some public commitment to the project before it is ever raised as a development control issue; indeed in the majority of cases public bodies have actually planned and financed the projects. Reservoir planning is the responsibility of the Regional Water Authorities and even in strictly commercial projects, such as the construction of golf courses, local authorities have normally been consulted previously and have a commitment in principle at least. There could be a serious conflict of local interest, but in practice the decision is normally put beyond local control, either because the proposed development contravenes the agreed Development Plan and is therefore 'called in' by the Minister, or because a public inquiry is held as a result of public opposition to the proposals. By their very nature major development proposals attract public interest and invariably organised opposition from existing and *ad hoc* pressure groups. Trial and error since 1947 have taught such groups that their best chance of a formal hearing is to persuade the Minister to 'call in' the proposal and to hold a public inquiry to help him make up his mind. Although there is no formal requirement for the Minister to accede to this demand, the public inquiry has become the accepted way in which the arguments for and against any major development proposals may be aired.

Road schemes are a category of development almost entirely outside the control of local planning authorities. Central government publishes its proposals, both in principle and in detail and there is no need for it even to seek approval prior to the commencement of building. Here again, however, public pressure has ensured that the procedure for considering and approving new road schemes is very similar to that for other development projects. When a proposal is put forward, the Department of Transport seeks the views of all sections of the community, including the local planning authority. If there is widespread disagreement, as there usually is over such complex issues, then the Minister has the power to order a public inquiry to help him make up his mind. It is very difficult to judge the impact of public inquiries on the outcome of major development issues because the decision-making process is never revealed. It is known that the inspector conducting an inquiry compiles a detailed report of the evidence and makes recommendations, but the Minister is not bound to accept them. The report of the inquiry is just one of the things that he must take into account in coming to a decision. Once he has made up his mind, his decision is final and although he usually explains his reasons, he is under no obligation to do so and the actual reasons for the decision may never be revealed (Levin 1979).

In the majority of cases the Minister confirms the development, not surprisingly in view of the fact that the bulk of proposals are initiated by government or government agencies in the first place. Nevertheless, there are a number of celebrated instances where the Minster has changed his mind, sometimes contrary to the recommendations of the public inquiry. The Roskill Commission that inquired into the relative merits of three alternative sites for a Third London Airport, came down decisively in favour of a site at Cublington in rural Buckinghamshire, but the Minister did not accept this, opting instead for another of the sites at Maplin Sands on the Essex marshes. Although the whole project was abandoned in 1974, further sites were later examined and Stansted airport was chosen in 1980. Leat has looked in detail at a much smaller proposal to construct two golf courses on an area of heathland at Woodbury Common, near Exeter in south-east Devon. Here again the report on the inquiry came down decisively in favour of allowing the development to proceed, but the Minister disagreed and turned down the application (Leat 1978). In both instances public opposition was extensive, well-organised and effectively orchestrated. At Cublington the Wing Airport Resistance Association had 61 766 members and was led by highly trained professionals, such as Evelyn de Rothschild, the merchant banker. It was able to put its case over most effectively through the media and individual members were in a position to lobby government directly (Perman 1973). In the Woodbury Common case too, careful presentation of the arguments against the golf courses through press and public meetings enlisted widespread popular sympathy and produced a petition with 15 000 signatures opposing the scheme (Leat 1978). In both cases the extent of the local opposition was clearly identified and it would appear that this both impressed the Minister and influenced him in coming to his decision. However, such a conclusion can be no more than speculation since in neither case did the Minister give a detailed explanation of how he came to his decision.

In the majority of cases the outcome and purpose of the public inquiry is less controversial, but equally important. Inevitably any major development will cause some local inconvenience and hardship and it is extremely difficult to foresee all the implications for small minorities. The public consultation process in general and the public inquiry in particular allow these problems to be identified and many can be overcome by quite minor adjustments to the original proposal. In another example, Leat has shown how it was possible to meet many of the objections to the A30 dual carriageway improvement between Exeter and Okehampton by tree screening and the provision of extra bridges over the new road (Leat 1978). From interviews he conducted, it was clear that the Road Construction Unit saw this as the main benefit of the whole public participation exercise. Questioning the validity of the whole project was officially considered at that time to be beyond the terms of reference of the general public debate, a view endorsed by Thomson who sees the indispensable role of public participation as that of resolving minority conflicts (Thomson 1978). In other words, despite the fact that the goverment has now accepted that the need for a new road, or an airport can be questioned at a public inquiry, it is still reluctant to use such a forum for policy-making. Only when the opposition is overwhelming, as it was at Cublington and Woodbury Common, does it fundamentally change its mind. If the public wish to exert direct influence on the number and size of major development projects, then they are still normally thrown back on more established democratic channels, rather than being able to question specific proposals when they are made.

Conclusions

This chapter has been concerned with the effects of the planning system on
some aspects of the development of rural settlement. The intention was not
to consider rural settlement change as whole. As a result many potent
influences, some of them government controlled, have been omitted. There
is, for instance, no discussion of the influence of the Development
Commission and its agency the Council for Small Industries in Rural Areas,
because it was felt that important as they were, they were not strictly part of
the land use planning system as such.

From the foregoing analysis it is clear that the planning system has
enjoyed a mixture of success and failure. There seems little doubt but that
the spread of urban areas has been sharply confined and that in the
countryside itself, new development has been concentrated largely within
existing settlements, rather than being spread haphazardly wherever
demand led. These are both major achievements and have contributed
greatly to maintaining a feeling of space in a densely populated country.

What the planning system has not done is to reverse the general decline in
the dispersed and self-contained settlement pattern that used to typify rural
Britain. The rural population has become increasingly dependent on urban
goods and services and at the same time access to them for many people in
country areas has become progressively more difficult. The planning system
certainly should have been in a position to ease such difficulties, but there is
little evidence that it has been able to do so effectively. It has even had only
limited success in mounting a cosmetic operation to preserve local building
styles and materials. Both the design and materials of rural housing are now
little different from that in urban areas, even in protected landscapes such
as National Parks. Nevertheless, despite these reservations, the rural
landscape would look very different had there been no land use planning in
the past 30 years, though the rural way of life would probably have been
much the same (Blacksell 1979).

7 *Management in the countryside*

Introduction

So far the emphasis has been on negative controls rather than positive management and while this is a fair reflection of the actual nature of most local government intervention, it exposes the severe constraints under which planning authorities have to work. Negative controls are permissive in that they allow authorities to arbitrate about the desirability of new developments if and when they are applied, but they can do little or nothing to initiate proposals. It also means that the authorities are very constrained in the extent to which they can regulate the nature and intensity of use, a point illustrated by the growing tendency for rural housing to be used for second and retirement homes, especially in the more attractive parts of the country.

Second and retirement homes

Second homes have been defined by Dower as, 'a residence which is exclusively or mainly occupied by someone who regards it as other than his primary residence' (Dower 1977, p. 155). On this basis it has been estimated that there are 350 000 (200 000 dwellings and 150 000 static caravans) in England and Wales (Bielckus *et al.* 1972) with a further 34 000 in Scotland (17 000 dwellings and 17 000 static caravans) (Aitken 1977). Although data are very scanty, the great majority of this new type of household has probably been created in the past two decades, with rapid growth occurring between 1960 and 1970 (Fig. 7.1). Since then there appears to have been a marked slowing down in the number of new second homes, in the short-term as a result of the economic situation in Britain, though in the longer-term reflecting a growing shortage of suitable properties (Dower 1977). Certainly in comparison with other European countries and North America, the proportion of second homes per head is still quite low. The rate could rise substantially in the 1980s, however, for the Housing Act 1980 gives sitting tenants the right to buy council houses, and in attractive areas there will be every incentive for those tenants to then sell their properties as second homes. The Act includes safeguards to prevent this occurring in the most attractive or preserved landscapes.

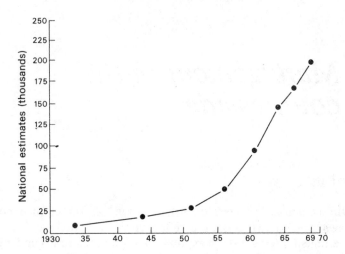

Figure 7.1 Estimated number of built second homes in England and Wales 1935–69. (Coppock, J. T. 1977. *Second homes: curse or blessing*. London: Pergamon.)

There is always a danger of over-stating the size of the problem posed by second homes (Coppock 1977). The most trouble is caused in those previously remote areas, where suspicion and resentment are aroused, and planners are under pressure to do something to stem the tide. In fact second homes comprise no more than 1% of the national building stock and only exceptionally, in parts of northern Wales where admittedly pressure has been locally considerable, does the proportion anywhere near approach the level of 30%. Indeed, in Wales the winter of 1979–80 saw a campaign of arson against second homes in which nearly 50 houses were destroyed by fire. A more typical distribution for a rural area is given in the report on the subject by the South-West Economic Planning Council, which tried to assess the extent of second home ownership for the whole of the South-West region (South-West Economic Planning Council 1975a). As can be seen from Figure 7.2, second homes were widely distributed in both Devon and Cornwall, but only in the extreme north-west of the latter county was more that 10% of the total building stock used for this purpose, for the most part the proportions were between 1 and 5%. A persistent problem with second homes that often leads to their importance being over-emphasised is the tendency for them to be grouped together in specific locations. A high proportion is concentrated along the coast, as in the Plymouth region and the South Hams in Devon. In Wales there also seems to be a smaller concentration in the less prosperous and remoter parts of the countryside, notably in Snowdonia, reflecting the lower property prices and the availability of vacant dwellings. Even so there is little indication that second homes are restricted to such locations. In England and Wales generally they are just as likely to be sited within existing settlements as outside them, save

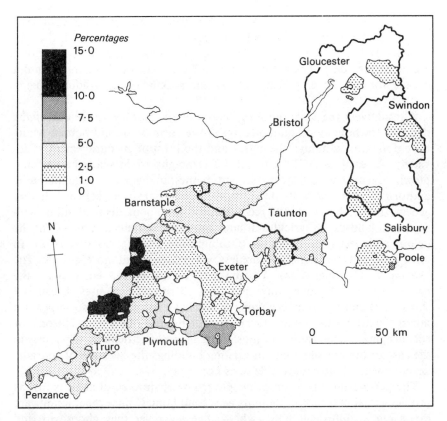

Figure 7.2 Second homes in the South-West as a proportion of dwelling stock. (South-West Economic Planning Council 1975. *Survey of second homes in the South-West*. London: HMSO.)

that where the locally economy was thriving, second homes were the first to be pushed out (Bielckus *et al.* 1972).

As far as control is concerned the major question is to what extent planners can influence the distribution and spread of this phenomenon. In effect the only control they can exercise over the use of dwellings as second homes is through the policing of new development, since a change from permanent to temporary occupation does not legally constitute a change of use. For many local authorities in the more attractive parts of the country the lack of direct powers is a source of considerable frustration. In the Lake District, for example, the Special Planning Board is trying to operate a non-statutory policy using Section 52 agreements whereby planning permission is only granted to applicants producing evidence that they are going to live and work permanently in the National Park.

Much greater powers exist in respect of caravans and a number of authorities have successfully restricted them to specific sites. Static caravans require a temporary planning permission and any that violate the terms of the published policy can gradually be removed as the permissions run out. An excellent example of the kind of success that can be achieved is the way in which Devon County Council has refused to allow static caravans within a mile of the coast. The policy has been rigidly enforced for more than a decade for new development and existing anomalies are gradually being cleared.

Alternatively the local authority can make a positive virtue out of the demand for holiday accommodation in the form of second homes. Many have done this, although the extent and the form of encouragement varies widely. As early as 1972 the county of Denbighshire in Wales had a fully-blown seven-point policy on second homes to help the local planning authority in coming to its decisions. The guidelines were couched in very general terms, but included recommendations about detailed siting with regard to landscape, agricultural land and other settlements, as well as the layout and design of buildings (Denbighshire County Council 1972). In most other authorities policies are less wide ranging, but more specific. The Dartmoor National Park Committee, for instance, encourages the conversion of redundant farm buildings to recreational uses, because it saves the building from becoming a derelict eyesore and also imposes a minimal extra burden on local services, which is one of the main reasons for encouraging concentration for most other forms of settlement. Obviously in the case of the redeployment of existing buildings the question of sporadic development in the countryside does not apply.

The permanent migration of people from urban to rural areas is just as problematical in terms of planning as second homes, since the migrants are exercising a quite legitimate right to live wherever they choose to find accommodation. However, their distribution means that locally they have radically altered the population structure and created demands on services quite different from those produced by a balanced resident population.

Part of the public apprehension arises from the way in which the statistics have been interpreted. On the one hand, Law and Warnes pointed to increases in the numbers of retireds between the 1961 and 1971 censuses of 32·6% on the north Wales coast, 32·1% on the Sussex coast, 28·1% in the Fylde, 26·9% in Lancaster, 25·5% in south Devon, 25·3% in Cornwall and north Devon and 25·2% in Kent, as compared with the national average of 19·2% (Law & Warnes 1975). They argued that if these trends were projected into the future, they were going to produce radical changes in the population structure with serious and expensive implications for the provision of services.

On the other hand, Gordon, in a study for the South-West Economic Planning Council, took a rather more optimistic stance (South-West Economic Planning Council 1975b). Using the concept of 'additional retireds', meaning the retired people in the population over and above those that would have been expected due to natural increase, he demonstrated that even though there was an increase in eight out of the nine regions in the

Table 7.1 Geographical distribution of the additional retireds: South-West Region 1971. (South-West Economic Planning Council 1975. *Retirement to the South-West.* London: HMSO.)

	Additional retireds ('000)	Total population ('000)	Proportion of additional retireds in the total population
North Gloucestershire	1	465	0·2
Bristol Severnside	15	930	1·6
North Wiltshire	−7	345	NA
Wellington Westbury	3	260	1·2
South-East Area	25	570	4·4
Exeter Torbay	51	465	11·0
Plymouth Area	16	345	4·6
West Cornwall	17	265	6·4
Bodmin Exmoor	17	215	7·9

South-West, only in Exeter–Torbay did it exceed 10% (Table 7.1). In contrast to Law and Warnes, he argued that the newcomers brought with them considerable additional financial resources and yet imposed a below average *per capita* burden on local authority services, in view of the fact that education consumes the lion's share of the rates. Arguing along these lines, a retirement option with major growth along the south coast was an option put forward at the draft stage of the Devon Structure Plan. Nevertheless, a proportion of the elderly also tends to be widely scattered in small picturesque villages and isolated country cottages, which poses logistical difficulties for medical and social services. Financial independence also makes these people less amenable to concentration into key settlements and the like, as has been attempted with families of school-aged children. Finally, although the specific demands of the elderly on services may be relatively small, local authorities still need to provide schools and recreation facilities in all areas, so that some of the savings may be more apparent than real. Nevertheless, Gordon makes a strong case for arguing that local opposition may be based more on emotion than reason. The economic objections are frequently over-stated and there is a refusal to adjust planning policies quickly enough to the new population structure.

Approaches to rural land management

The frustrations caused by the ineffectiveness of planning controls for influencing these and other forms of countryside change have led to direct intervention not only being talked about, but practised on a growing scale. However, the extent to which public authorities, at both national and local level, should attempt to manage land and development directly is still far from clear. Some early advocates of land use planning thought that planners should assume a position that enabled them to orchestrate events, not merely to react to them. For instance, Dudley Stamp, Director of the First Land Utilisation Survey of England and Wales in the 1930s and vice-chairman of the 1942 Scott Committee still believed in 1961, that 'The planner's task is to determine the optimum use, in the national interest, of every acre of the surface. Indeed his task goes further. Many acres of land serve more than one purpose, as when hill moorland is used as a gathering ground for a water supply, pasture for sheep and is also open for access for walkers seeking air and exercise. Multiple use of land must be promoted whenever possible' (Stamp 1961, p. 65).

Support for this view, based on the allocation of land by planners, has continued over the years in a number of different guises (Eversley 1973), but at the same time has also attracted fierce criticism. A proposal by Tranter that land use planning quotas be determined nationally and then transmitted to the regional and local levels through the administrative hierarchy has been condemned as being anti-democratic, costly both to introduce and run and too narrowly conceived (Tranter 1978, Whitby & Thomson 1979). In many ways the last point is the most fundamental, for it questions whether it is possible, or desirable for government to predetermine land use priorities. Whatever the physical attributes of a particular site, prevailing economic and social circumstances should be just as important for deciding how the land may best be used and both are prone to change rapidly. The argument is thus one of long-term physical planning based on resources against short-term economic exploitation.

One basic rural resource is agricultural land and it has often been argued that suitability for agriculture should be the deciding factor in allocating land, but land use capability has so far played only a minor role in determining planning policy. For more than 20 years the Soil Surveys of

Table 7.2 Land Use Capability Classification for England and Wales (Bibby, J. S. and D. Mackney 1969. *Land use capability classification.* London: Soil Survey.)

Class 1	land with very minor or no physical limitations
Class 2	land with very minor limitations that reduce the choice of crops and interfere with cultivations
Class 3	land with moderate limitations that restrict the choice of crops and/or demand careful management
Class 4	land with moderately severe limitations that restrict the choice of crops and/or require very careful management practices
Class 5	land with severe limitations that restrict use to pasture, forestry and recreation
Class 6	land with very severe limitations that restrict use to rough grazing, forestry and recreation
Class 7	land with extremely severe limitations that cannot be rectified

Each category may be further subdivided by the following sub-class limitations: c. climate, e. erosion, g. gradient or soil pattern, s. soil, w. wetness.

England and Wales and Scotland have been preparing Land Use Capability maps based on soil profile characteristics, which divide land into seven categories according to its suitability for agriculture (Table 7.2). One of the stated aims of the surveys is to provide detailed information for land use planning but while the maps have been useful for establishing general strategic guidelines, they are not sufficiently accurate to be of much help in determining whether or not individual developments should be allowed (Mackney 1974). This limitation is also found in the Land Classification maps prepared by MAFF, which divide land into five grades (MAFF 1977a). Indeed the DOE has advised local authorities that both sets of maps 'are not suitable for use in evaluating individual sites' (DOE 1976a). Boddington has also criticised the MAFF maps for ignoring completely likely financial returns, which can be shown to vary widely, even when the soils have very similar physical characteristics (Boddington 1978). In other words, even if a national policy for land use based on agricultural land quality were deemed to be desirable, it would almost certainly prove impossible to provide the necessary detailed data. Not only are the physical assessments of land potential still too crude for them to be of much help in determining planning applications, there is also widespread disagreement about the extent to which socio-economic criteria can, or should be included in land capability assessments.

It is not, however, just the lack of a reliable and generally accepted method for assessing the relative merits of rural land for any use that has prevented the adoption of more positive development strategies. Professional planners have found it difficult to counter the argument that policies designed to achieve one end have in fact produced the opposite effect (Jacobs 1978). All too often the release of building land around villages has only encouraged housing for commuters or holiday development. New factories and warehouses, rather than being praised as important initiatives in the fight to halt rural decay, have frequently been criticised as ugly and inappropriate intrusions into the rural scene.

Nevertheless, the belief that planners ought to be adopting a more positive role in the management and use made of the countryside still persists, even amongst the most hard-bitten members of the profession. One area where more positive management proposals have been tried are the ten National Parks that were created between 1950 and 1955 and in which there was a clear invitation to public authorities to adopt a more interventionist note in the national interest, although neither the funds nor the manpower were made available for this to be a practical possibility in any comprehensive fashion for the first 25 years from 1950 to 1974. Even so, the Sandford Committee that reviewed the successes and failures of the National Parks in 1973, was able to report that, 'the parks are now almost universally accepted; public concern for their protection has become strong and widespread. There is general approval of the statutory purposes of the parks – the preservation and enhancement of natural beauty (which embraces scenic beauty and wildlife), and the promotion of their enjoyment by the public' (DOE 1974d, p. 8). The committee went on to recommend that the authorities responsible for the National Parks should approach their management functions, particularly in relation to traffic regulation and provision for recreation, in a much more direct and positive manner.

This more positive approach had already been advocated in 1967 (Hookway 1967) and the need for an overall agency, charged with the welfare of the countryside, was answered in 1968 by the Countryside Act that replaced the National Parks Commission by the Countryside Commission, a recruit covering all rural areas and not just those deemed worthy of special protection. Section 11 of the Act also made it mandatory for all government departments to pay due attention to the desirability of conserving the natural beauty and amenity of the countryside and also of developing its potential for nature conservation and recreation. Funds have subsequently been made available, mostly through the Countryside Commission, to help both public and private bodies to make wider use of rural land and to accommodate the varied pressures to which it is exposed. For example, the Act gave local authorities the powers to purchase land, compulsorily if necessary, for the purposes of setting up Country Parks to satisfy the demand for open-air recreation and also provided up to 50% Exchequer grant aid (75% for private individuals or organisations) to help with initial establishment costs.

Despite these improvements countryside management is still poorly co-ordinated; the initiative for management proposals has to come from the individual local authorities, although they can now expect financial support and general encouragement from central government if they decide to act. The situation has been much improved since the Local Government Act 1972 that established separate departments for each of the National Parks within the appropriate local authority and provided 75% of the necessary funds from the Exchequer. The form of these new departments was not specified save in two respects: each had to appoint a National Park Officer and to prepare a National Park Management Plan for submission to the Countryside Commission by April 1977. It was a landmark, for this was the first time that any public authority had been required to produce general land management proposals for the area under its control.

Whatever the merits of National Parks in England and Wales and the management structures that have been evolved for them, they have certainly attracted disproportionate attention and resources in relation to their size. In Scotland, fears about just such an imbalance were one of the main reasons why the proposal of the Ramsay Committee in 1945 that a system of National Parks be set up there was rejected (Cmd 6631 1945, Cmd 7814 1947). The price has, however, been a heavy one for throughout the 1950s and 1960s land use priorities in Scotland were not even discussed within the limited context of a national parks system and the way forward was determined by the relative strengths of all the interested parties, private land owners, local government and central government and its agencies like the Forestry Commission.

In an attempt to bring more order into the situation the Select Committee on Scottish Affairs decided in 1972 to investigate the whole question of land resource use in Scotland. Its report identified the main problems as a lack of clear guidance from central government and the limited perspectives of the official government agencies such as the Department of Agriculture for Scotland and the Forestry Commission. The solutions proposed were the setting up of a land use council to advise on land use priorities, a strengthening of the Countryside Commission for Scotland and the Nature Conservancy, and the drawing up of a National Structure Plan (House of Commons 1972).

The government welcomed the analysis, but rejected all the proposed solutions in favour of a more broadly based consultative body. The most important created so far is a standing committee with representatives from all the central government agencies with an interest in Scottish land use matters (the Department of Agriculture and Fisheries for Scotland, the Scottish Development Department, the Scottish Economic Planning Department, the Countryside Commission for Scotland, the Forestry Commission and the Nature Conservancy Council). The committee has met regularly since 1973 and produced some most useful guidance on land use planning matters in the form of factual Land Use Summary Sheets and Planning Advice Notes. It has not, however, provided the kind of broadly based forum needed to agree a national strategy for land use. Meetings are private and the membership is restricted entirely to government departments or agencies (Coppock 1979). As a result Scotland, like England, still lacks any clearly defined strategy and has only avoided serious land use conflicts because until recently the demographic and general economic pressures on much of its land surface have been slight in comparison with those found in England and Wales.

Despite the failures in all three countries to reach agreement about national land use strategies, the Countryside Commissions have both striven hard to fill the gap. In England and Wales the Commission has concentrated on extending the work of its predecessor beyond the National Parks and the uplands in general, where most of the attention has been concentrated, into the agriculturally more intensively used lowlands. On the one hand the work of designating tracts of countryside as either National Parks or AsONB has continued and although no new National Parks have emerged, ten more AsONB have been designated and confirmed since 1968, seven of them (Norfolk Coast, Kent Downs, Suffolk Coast and Heaths, Dedham Vale, North Wessex Downs, Mendip Hills and the Lincolnshire Wolds) in lowland England.

Work has also begun on some new approaches to rural land management, most of which originated in the *New agricultural landscapes* study (Westmacott & Worthington 1974). The findings of the report, which are discussed in Chapter 4, have been instrumental in providing the inspiration for some of the most important recent initiatives by the Countryside Commission. For example, a considerable effort has been put into amenity tree planting schemes and, encouraged no doubt by the widespread national concern about the ravages of Dutch elm and other diseases affecting the country's ageing tree stocks, there has been a good response. On a wider scale, Demonstration Farms have been set up to show farmers the management implications and possibilities of landscape conservation in agriculture. Most important of all, however, are the New Agricultural Landscape Projects. Two of these were set up in 1977 in Hereford & Worcester and Suffolk and a third is to commence in 1979 in Cambridgeshire. The area covered by each is about 5000 ha and the aim is to appoint a project officer to advise all land owners on possible sources of grant aid for landscape improvement schemes. He also has a special budget for mobilising voluntary effort, a potent weapon in the fight to conserve the rural landscape but one which has often proved difficult to harness. Even if these actions amount to something less than a national strategy, they at least represent a commitment to orderly change within an overall framework, which includes all the countryside and not just those areas deemed worthy of special protection.

In Scotland, the Countryside Commission has approached land management from a different direction based on the more traditional use of designating certain areas for specific uses. Their alternative has been to develop two separate systems, defining scenic areas to encompass the needs of landscape conservation and a hierarchy of parks specifically for recreation. The first category is the Urban Park that, as its name suggests, is essentially local in character, administered by the local authority and catering for the immediate daily recreational needs of the country's towns and cities. Next comes the Country Park, where it is intended that relatively large areas of countryside be set aside for the public to enjoy outdoor recreation pursuits. It is important to note that in both the Urban and the Country Parks the emphasis is firmly on accessibility, rather than the absolute quality of the scenery (Countryside Commission for Scotland 1974).

The Regional and Special Parks, the two upper tiers in the system, are described in more speculative terms, not least because they relate to future developments, rather than existing facilities. They also differ markedly in form. Whereas the land in the Urban and Country Parks is used exclusively for recreation and probably owned or leased by the local authority, the higher level parks will be designations similar to the National Parks and AsONB in England and Wales, where recreation is but one of a number of uses.

Regional Parks do already exist − the Clyde Muirshiel Regional Park in Strathclyde and the Lomond Hills Regional Park in Fife − but their precise function and status within the park system is still a matter of some debate.

As far as the Commission is concerned they should serve two purposes. In the short-term they should provide comprehensive facilities to satisfy immediate demands for a variety of recreational uses, but they should also act as reservoirs of land for accommodating future recreational needs. This type of Park is particularly appropriate to Scotland, where the regional authorities are such an important element of the local government structure.

The final category, the Special Parks, will be expected to help satisfy the national and even international demand for recreation in Scotland, especially in the most beautiful areas of the country. The pressure on the Cairngorms, Glen Nevis, Glen Coe, Loch Lomond, the Trossachs and other areas is intense and growing and, however much the Countryside Commission would like to separate landscape conservation and recreation, they have to recognise that the landscape itself is a recreational resource. These Special Parks go some way towards admitting this case and, inevitably, bring the planners face to face with the frustration of trying to reconcile the irreconcilable, which has been such a thorn in the flesh of the National Parks in England and Wales. Nevertheless, the whole purpose of the Special Parks concept would be to limit the areal scope of the conflict, by intensive management planning in careful selected locations.

Having established the guidelines for recreation, the Commission published its plans for landscape conservation in 1978 (Countryside Commission for Scotland 1978). It identified forty national scenic areas across the country, covering 12·7% of the total land area. The designations are unique because they are based solely on a careful assessment of landscape quality, but they are only a beginning as they include no proposals for direct management. The areas were confirmed in 1980 and it was announced that extra development control powers would be available in them. Under the new procedure, all but modest developments will have to be referred to the Countryside Commission for Scotland.

The indecisiveness that has characterised the debate about approaches to rural land management and the partial nature of those management strategies that have been implemented mean that the record of achievement is inevitably a fragmentary one. Central and local government authorities have opted for a wide variety of solutions at different places at different times and there is little internal consistency in the strategies overall. Any assessment of what has been achieved will inevitably reflect this. The rest of this chapter first looks at some of the administrative structures that have been created for management purposes and then proceeds to consider the effectiveness of some of the more important management techniques they have employed.

Rural development boards

The principles of regional policy have been broadly accepted in the United

Kingdom for nearly half a century. Successive governments have pursued policies to compensate remote and economically disadvantaged areas for the accident of their relative location, so that social benefits may be more or less equally distributed. Despite their concern for achieving spatial uniformity, these policies have not been concerned with land use *per se*, but have concentrated on encouraging industrial development through various incentive schemes. More specific aid has been provided by the Development Commission and by grants to agriculture, particularly from the European Community to aid what it calls 'less favoured areas'.

In the most remote areas, however, this sort of approach has proved to be largely inadequate and has had to be supplemented by the creation of development boards, responsible for promoting the general advancement of specific regions. The first, and by far the most successful of these is the Highlands and Islands Development Board that was set up under the Highlands and Islands Development (Scotland) Act 1965. Its area, which is shown in Figure 7.3, encompasses about half the area of Scotland but only about 322 000 people, the vast majority of them widely scattered in small communities of less than 10 000 people. The Board's main duties are to

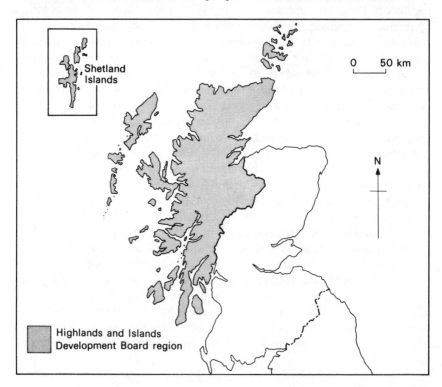

Figure 7.3 Area covered by the Highlands and Islands Development Board. (Tranter, R. B. (ed.) 1978. *The future of upland Britain*. Reading: Centre for Agricultural Strategy.)

assist the inhabitants in improving their economic and social conditions and to enable them to play a more effective part in the development of the nation. The long-term objective is to reverse the steady out-migration from the region by revitalising agriculture and other traditional industries and by encouraging the introduction of new forms of economic activity (Farquarson 1978).

To accomplish these tasks the Board has a budget of £11 million (1976) and there have been three main thrusts to the work. First, it has adopted a supportive role: provided assistance with capital funding for commercial initiatives; built new factories and other facilities to attract new investment; and also provided retraining, so that the predominantly agricultural population will have the necessary skills to take advantage of industrial opportunities. Secondly, the Board has begun to foster new development on its own initiative. Traditional skills such as knitting and weaving, in danger of disappearing from the commercial scene through lack of production and marketing expertise, have begun to be revitalised through the Board itself providing the commercial framework. Thirdly, the Board plays a very important role in representing the interests of the Highlands and Islands in central government. This involves development on a much larger scale than that required for rural regeneration, but the Board sees no reason why the region should not have the opportunity of enjoying the industrial benefits of North Sea oil and gas, for example, rather than accepting the crumbs that fall its way as the consequence of development elsewhere.

In many ways the work of the Highlands and Islands Development Board is in the main stream of regional policy in the United Kingdom but there is one crucial difference, which justifies its inclusion in a discussion of rural land management. Although it sees its function as to plan for and to encourage change, it does not want this to proceed in such a way or on such a scale that it will damage the social balance or unique environment of the area. In other words the main commitment is to the region as an entity and the Board will use appropriate means to foster its rehabilitation as long as these do not threaten its basic characteristics (Alexander 1978).

It is always difficult to gauge the success of such a body, but by common consent it has identified many of the root causes of the economic decline in the region and set in train some useful remedial action. It has also helped to give that part of Scotland an identity of its own. For all these reasons governments have been encouraged to try to repeat the experiment elsewhere. In 1968, under Section 45 of the Agriculture Act 1967, the North Pennines Rural Development Board was established, although within three years it was wound up. It had failed to generate the kind of cohesion and enthusiasm of its Scottish counterpart, for what reasons is not clear but two important differences may provide clues. The brief of the North Pennines Board was very much narrower, being confined almost entirely to agriculture, thus precluding it from looking at other sources of development. Furthermore, it did not have the same commitment to halting rural depopulation and, as a result, may well have forfeited the kind of public goodwill which has helped sustain the Highlands and Islands Development Board (Capstick 1978).

In 1976 another type of body, the Development Board for Rural Wales, was established. This was developed from a succession of earlier bodies concerned with rural regeneration in central Wales, each of which was superseded because its brief was too narrow, being concerned either with industry or agriculture, but not both. The new Board has this broader brief and may in time, therefore, emulate the achievements of the Highlands and Islands Development Board, but as yet it is too early to make any judgement one way or the other.

National Park Plans

The ten English and Welsh National Parks offices are the orgainsiations that probably come closest to the ideal of multiple rural land management. When their administration was reorganised in 1974 each was required to submit to the Countryside Commission a 5 year management plan for the parks as a whole by April 1977. By the end of 1977 all had done so and the plans have been described by the Director of the Countryside Commission as 'a milestone in the development of planning' (Hookway 1978).

Although the form and detailed structure of the National Park Plans vary, there is remarkable uniformity in the analysis of the problems facing the National Parks and in the solutions proposed. All strongly reiterate the twin statutory purposes of landscape conservation and the provision of opportunities for open air recreation. They also accept the recommendations of the Sandford Committee and DOE Circular 4/76 that where the two objectives are in conflict landscape conservation should take precedence, and that the only secure basis for landscape conservation is a thriving and economically prosperous local community. It follows that the management objectives adopted by committees may be broadly grouped under three headings; conservation, recreation and community affairs. A commitment to widespread consultation with all appropriate local authorities and with other *ad hoc* interested parties, such as amenity societies is also universal, and all have proposals for the internal organisation of the National Park departments and the services they offer (Dennier 1978, 1980).

Undoubtedly the Plans have gone some way towards making multiple land use management a reality, but their successes must be kept in perspective. In almost every instance where there has been pressure for major change, either through enclosure or agricultural improvement, or through large-scale development proposals, such as an application to extract minerals, build a reservoir or a power station, or build a road, the National Park purpose has been overriden and all the fine plans have proved worth little more than the paper they were written on. A start has been made on agreeing strategies for multiple land use in these areas, but the agreement is a fragile one and under persistent threat.

Management agreements

A major problem in many upland areas and particularly in National Parks is to persuade private landowners to manage their land in accordance with National Park and other planning objectives. Many parts of the rural economy, and notably farming and forestry, are almost completely outside planning control, so that an alternative framework has to be devised if irreconcilable conflicts of interest are to be avoided. One of the most obvious is the management agreement, whereby a landowner voluntarily enters into written agreement with the authority to manage his or her land in a particular way. In one form or another this device has been available since Section 34 of the Town and Country Planning Act, 1932 first put it on the statute book, but has only been widely used since the National Parks and Access to the Countryside Act 1949 came into force, when specific duties were laid upon those in charge of National Parks to look after rural land. The precise form of the agreements varies widely depending on the particular circumstances, but the financial arrangements may be roughly divided into three categories. At one end of the scale the agreement may have no financial implications and simply be an agreement by the landowner with the National Parks authority to manage land after a particular fashion. At the other extreme the agreement may attempt to take account of the full opportunity costs of adopting the agreed management strategy. Both are unusual and the most common solution is for the landowner to be paid some token compensation for agreeing to his freedom of action being curtailed. Not surprisingly the rather vague guidance on the financial arrangements has been the most frequent cause of negotiations on management agreements breaking down (Feist 1979). There are, however, other problems. If an agreement is violated, enforcement is extremely difficult both for practical legal reasons and because any protracted court action would seriously jeopardise future management agreements. Landowners have also proved unwilling to tie themselves down indefinitely, or even for any extended length of time for fear of reducing the potential value of their holding. Finally, there is considerable uncertainty as to whether many existing agreements can be interpreted as remaining with the land if the title passes to another owner.

Despite these difficulties, some National Park Committees have set great store by management agreements. In Exmoor, where the remaining stretches of moorland are seriously threatened by ploughing, management agreements are currently being used in the absence of compulsory powers. In 1979 the outgoing Labour government tried to pass legislation enabling

Moorland Conservation Orders to be declared by the National Park Committee, which would prohibit conversion, following the advice of the 1977 Porchester report (DOE 1977a). However, before the bill could become law, the government fell and the succeeding Conservative administration has declared a preference for voluntary agreements rather than any form of compulsion. Unfortunately, the number of farmers prepared to enter into management agreements for the purposes of limiting the conversion of moorland by ploughing is small in comparison with the size of the total area at risk.

To a large extent this is a general verdict on management agreements, which although welcome in their own right, are invariably on too small a scale to be useful as a comprehensive means of control.

The Upland Management Experiment

Recognition of the need for a broader and more all-embracing type of approach led the Countryside Commission to establish and fund, in conjunction with the Lake District Planning Board and the Ministry of Agriculture, Fisheries and Food, the Upland Management Experiment in 1969 covering a small area of the Lake District National Park to the south and east of Ullswater. The aims of the experiment were:

(a) To test a method of reconciling the interests of farmers and visitors in the uplands by offering financial encouragement to farmers to carry out small schemes that improve the appearance of the landscape and enhance the recreational opportunities of the area;
(b) To assess what effect, if any, this method will have on farmers' attitudes towards recreation and landscape (Countryside Commission 1976).

The experiment was a considerable success. Largely due to the efforts of the project officer, who carefully explained to farmers and other local residents how they might benefit from the attractiveness of their local environment for tourism and recreation. As a result more positive attitudes towards both the landscape and visitors began to develop. Features such as dry stone walls and traditional buildings, which had previously tended to be dismissed as relics of an earlier age, began to be valued as assets which attracted people and money to the Lake District. Nonetheless, many local people remained sceptical. In 1973 the experiment was extended to cover nearly a third of the National Park, but although the approach was again generally welcomed, the whole initiative became somewhat bogged down. The main problem was the lack of success in finding a way of translating the new-found local appreciation of the benefits of the traditional landscape for generating income from tourism and recreation into more formal management agreements. To try to find a way out of the impasse the reconstituted and renamed Lake District Special Planning Board commissioned a very detailed study of a small part of the Park, stretching from the village of Patterdale up the valley to the village of Hartsop and Brothers Water and then on to the summit of the Kirkstone Pass. The whole area was christened by the study team, the Hartsop Valley, and their investigation made a number of detailed recommendations about how the Upland Management Experiment might be revitalised. The main recommendations were that there ought to be better co-ordination among the advisory and grant giving agencies in government, preferably using the good offices of the Upland Management Experiment; that the Board ought to take a more positive stance on those management issues over which it did exercise control, for example giving planning permission for more tourist accommodation and maintaining footpaths; and that there ought to be better compensation and insurance for local residents against damage caused by tourists and that this should be the responsibility of the Board (Feist, Leat & Wibberley 1976).

It is still too early to say whether these recommendations will actually succeed in making the experiment a central management tool. The Lake District Special Planning Board feel confident enough to have changed its status from an experiment and to rename the approach the Upland Management Scheme. Elsewhere it has been noted that all the other National Parks have to some extent adopted the ideas developed during the experiment as part of their strategies for upland management (Dennier 1978). If the promise is fulfilled, it will have been a most significant breakthrough in the difficult task of exerting a measure of public control over the management of rural land, other than through the medium of development control, and certainly represents a positive approach.

The Dartmoor Commons Bill

There are large areas of common land in a number of National Parks and they present a tantalising management challenge. The main reason for their existence, of course, is to provide grazing land for the commoners' animals, but because the owners and the commoners are by no means always one and the same and because many of those with common rights now barely exercise them, the agricultural management of these areas has often deteriorated badly. On the other hand the existence of large tracts of open country with rough moorland vegetation has been an invitation to walkers to wander at will. Freedom of access has been assumed for many centuries, although anyone walking on common land without permission is a trespasser, unless a special agreement has been made.

Common land is a particularly live issue on Dartmoor where 38 000 ha (Fig. 7.4) or 41% of the National Park, are commons. In the 1970s

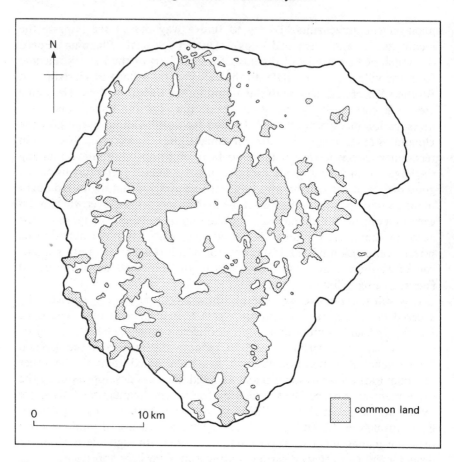

Figure 7.4 Common land in the Dartmoor National Park. (Dartmoor National Park Authority 1977. *The Dartmoor National Park Plan*. Exeter: Devon County Council.)

agricultural management had reached a very low ebb, despite the existence of the Dartmoor Commoners' Association that was formed in 1954, and the commoners themselves approached the National Park Committee with a view to their jointly promoting a private bill through parliament to regulate the use of the commons.

After four years of hard negotiation, the Dartmoor Commons Bill was deposited in November 1978. The bill's main concerns were to regulate agricultural use of common land and to provide a statutory right of access to the public on foot. Agricultural management was to be in the hands of a new Commoners' Council, the members of which would have been commoners, landowners and, significantly, two members of the National Park Committee. This was viewed as a breakthrough, for it was almost the first time that the farming community had recognised, evenly tacitly, that the Committee has a positive role to play in land management. The Council would have been responsible for regulating all stock on the commons and would also have had very limited powers for improving the vegetation in the interests of better animal husbandry. The National Park Committee would have been responsible for managing public access both on foot and on horseback, although the latter would not be a statutory right. To do this the Park Wardens would have had powers to close certain areas temporarily in the interests of better management and would also have been able to prohibit certain inappropriate activities, such as large gatherings and disruptive sports.

If the bill had succeeded, it would have provided the kind of coherent management for a most significant development in rural land management. For almost the first time controlled multiple use on a large scale would have been a reality. Although the proposals were specific to the Dartmoor commons, there is good reason for supposing that they could have been copied more generally if successful. Ever since the Royal Commission on Common Land and the Commons Registration Act which followed it in 1965, national legislation has been awaited, but the painfully slow progress on registration has repeatedly delayed further action (Cmnd 462 1958, Vane 1977). In 1978 however the government did issue a consultative document, so that the matter does now seem to be moving slowly towards a conclusion (DOE 1978a). If it had been successful, the Dartmoor Bill could have done much to speed up the process and many of its clauses might have been used as models in the national legislation. Unfortunately, it was for this very reason that the Bill fell in the House of Commons in April 1980.

Access agreements

The issue of public access to open country was at the heart of many of the land use conflicts in National Parks long before they were ever designated as such. The National Parks and Access to the Countryside Act 1949 made provision for access agreements over open country along similar lines to management agreements in both National Parks and AsONB. Subsequently the Countryside Act 1968 broadened the scope of the legislation to include woodlands, rivers and canals within the definition of open country and removed the necessity for designations to be confined to National Parks and AsONB. As can be seen from Figure 7.5, the area covered by access agreements and areas acquired by local authorities for access purposes has grown steadily and in 1975 stood at 31 238 ha (Gibbs & Whitby 1975). According to the Sandford Committee, agreements are generally welcomed by both landowners and the public and have been a most successful management tool (DOE 1974d).

Even so these agreements still pose problems. The same kinds of difficulty over financial compensation arise as with management agreements and often prevents negotiations being successfully concluded.

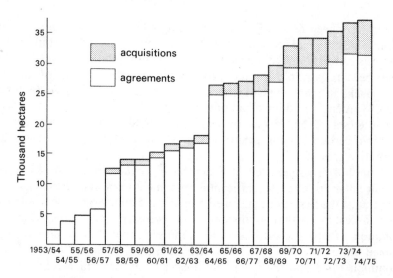

Figure 7.5 Number of access arrangements in England and Wales 1953/54–1974/75. (Gibbs, R. S. and M. C. Whitby. *Local authority expenditure on access land.* Newcastle upon Tyne: The University.)

There is also a certain wariness in a number of National Parks about the implications of access agreements in particular areas for public access as a whole. In areas such as the Dartmoor commons where the public enjoy *de facto* access without any kind of hindrance, access agreements are viewed with some suspicion, since many people feel that their main effect may be to limit freedom of movement rather than guaranteeing it. It is perhaps a wry comment on the Dartmoor Commons Bill, that its provisions for public access and visitor management were in fact an access agreement by another name, although this does not appear to have been recognised so far in the debate. In the future it seems likely that access agreements will increase in importance. As more land is enclosed and intensively managed, *de facto* access to open country will become less and less relevant as a general means of ensuring that the general public can enjoy informal recreation in the countryside and formal provision will become a virtual necessity.

The Urban Fringe Experiment

Evidence of the Countryside Commission's desire to broaden their field of interest beyond National Parks and the AsONB is provided by the proposed Urban Fringe Experiment. In an attempt to encourage more positive use of the large areas of rural land around the major urban areas, the Commission intends sponsoring a number of projects which would eliminate derelict land and intensify use, especially for recreation. The idea is based on a number of successful pilot schemes in the outer London boroughs of

Havering and Barnet and in the Bollin Valley in Cheshire, and it is proposed that the first site for the experiment proper should be the boroughs of St Helens and Knowsley in the Merseyside Metropolitan County between 1980 and 1985.

Other approaches

Countryside management is now coming to be viewed as essential in most rural areas, if open conflicts are to be avoided. Since the early 1970s, other bodies such as the Water Authorities and the Forestry Commission have been encouraged to expand their activities beyond the narrow service they were originally set up to provide. The Water and Forestry Acts 1973 both make it mandatory for authorities to develop their respective facilities for recreation. Local authorities with land other than nationally designated landscapes, have also been increasingly encouraged to foster positive countryside management.

The most obvious areas into which to extend countryside management practices were the AsONB. Although there is no statutory duty for these areas to be administered separately, the Countryside Commission has always strongly recommended that these areas should at least have advisory committees, or, where the area involved covers more than one authority, joint advisory committees. Somewhat surprisingly there has been little enthusiasm for the idea amongst local authorities. Only six AsONB have such committees – Chichester Harbour, the Chilterns, the Cotswolds, Dedham Vale, the Quantocks and the Wye Valley – although two others have similar but differently constituted bodies. In Arnside and Silverdale there is a joint committee of parish councils and in the Malvern Hills most of the AONB is watched over by a Board of Conservators.

Similar problems also affect the 44 stretches of Heritage Coast which have been, or are about to be defined in England and Wales. None is administered separately, although a few, notably the East Sussex Heritage Coast do have the services of a full-time project officer.

The limited scope for positive management in AsONB and Heritage Coasts has made the Countryside Commission look eagerly for opportunities to create new National Parks in those areas where the countryside is under severe pressure. The results of these efforts have not been encouraging. In 1972 the Secretary of State for Wales refused to confirm the order setting up the Cambrian Mountains National Park, after strong opposition from local farmers and residents, and in 1976 the Countryside Commission's proposal that the East Anglian Broads be made a National Park was withdrawn in the face of implacable local opposition (Countryside Commission 1979a).

In the Broads an alternative arrangement has now been adopted, which it is hoped will enable positive management of this area to proceed and to produce some resolution to the strong conflicts of interest between farmers, the tourist industry and the needs of nature conservation. A 26 man authority has been set up, drawn from the county and district councils, the relevant harbour authorities, the Regional Water Authority and with three members nominated by the Countryside Commission. The authority has appointed a Broads Officer and, financed from both local and national sources, he will have to produce a management plan (Blacksell 1978, O'Riordan 1979). The proposal is bold and innovative and there should be little difficulty in identifying the management problems in this area, but it remains to be seen whether the Broads Officer will be able to persuade the warring factions amongst the statutory authorities to accept his proposals for remedial action. For example, in 1979, the Anglian Water Authority proposed to build a flood barrier at Yarmouth, which would totally change the wetland nature of the Broads and transform meadowland into arable land.

Unquestionably the need for compromise and agreement between the many parties with a stake in the countryside is coming to be more widely accepted. The New Forest in Hampshire well illustrates the gradual change of heart. For nearly 1000 years this area has been managed as a commercial forest with subsidiary hunting and agricultural activities. Since 1922 the Forestry Commission has acted as the government's agents and had developed a clear policy for the future, but in the 1960s and 1970s the whole basis of their calculations began to be undermined. First, they underestimated the importance for nature conservation of the forest and were forced to reassess their plans for commercial forestry in the face of considerable public pressure. Secondly, they found that the recreation pressures (there were an estimated 800 000 campers alone in 1975) were growing so intense that positive management was essential if the visitors were not to destroy the unique habitat and the commercial future of the Forest. Accordingly in 1971 they set about a programme of careful management unobtrusively directing visitors to appropriate areas, providing basic facilities for camping and other outdoor pursuits, and publishing literature to improve the quality of countryside interpretation (Forestry Commission 1975). The results have been spectacular for the New Forest is now acknowledged as a model for successful management, where the interests of forestry, agriculture and recreation are successfully combined, and the Forestry Commission have broadened their horizons, so that they see themselves as managers of land in multiple use, rather than producers of a commercial timber crop (Small 1979).

Conclusions

The way in which the attitudes to positive management in the countryside have changed in recent years is most noticeable. Although there is still no national land use policy and approaches to multiple use vary widely, depending on the context in which they are initiated, there is now general agreement that positive intervention to promote more equitable and effective patterns of land use is a legimate function of government. Even local authorities in areas where there are no protected landscapes or chronic social and economic problems are beginning to accept the benefits of management and, like the urban areas before, the countryside itself will be the chief ultimate benefactor (Probert & Hamersley 1979).

8   *The way ahead: closer co-ordination?*

The widespread concern about the future of the British countryside is self-evident. Private misgivings have resulted in a steady stream of legislation over the past 50 years and a proliferation of official controls, both negative and positive, by central and local government. Even so, it is still far from clear to what extent all this activity has been part of a long-term strategy, or merely a string of *ad hoc* decisions taken to deal with specific crises as they arose.

In the 1930s there was clear if fragmentary evidence, of attempts by the government to enact comprehensive policies: legislation was introduced to deal with the control of development, to promote the revival of agriculture, and to provide for recreational access to open land. However, an official report at the end of the decade concluded that these powers were inadequate; the standard of design was generally poor and many buildings erected in the open countryside were inappropriately sited. The report also bemoaned the large amount of unnecessary destruction in the countryside (Health, Ministry of 1939).

Faced with the abundant evidence of the general rundown state of agriculture that could be seen by anyone venturing into rural areas, the government commissioned the 1942 Scott report, the only really comprehensive study of rural land use conducted this century (Cmd 6378 1942). It identified five major causes for concern:

(a) The lack of overall planning direction,
(b) The lack of industry and commerce,
(c) The poor state of agriculture,
(d) The poor state of village life,
(e) The need to conserve rural landscapes, while at the same time making them accessible.

The report stressed the need for an overall countryside planning strategy under a single, central agency, 'Unless the various types and units of development are coordinated and regulated, and, where necessary, prohibited by suitable machinery, the best use of the land of the country will not be attained. This is the true function of "planning" ' (Cmd 6378 1942,

p. 84). The case was, however, weakened by another recommendation. In spite of the evidence indicating that conflicts between agriculture and forestry and other land uses were at the root of many of the land use problems in the countryside, the Scott report advised that both should be excluded from planning control. The decision was of fundamental importance, for, as the evidence in this book has shown, the schism between planning for agriculture and forestry and land use planning in general has been one of the basic barriers to the emergence of a comprehensive policy for rural areas.

The establishment of a single agency for rural affairs was further hampered by the fact that the pressure of post-war parliamentary business made it impossible to introduce all the necessary legislation in one major Act. Instead there were more than half a dozen passed between 1946 and 1949, the most important being the Town and Country Planning Act 1947, the Agriculture Act 1947, the Forestry Acts 1945 and 1947 and the National Parks and Access to the Countryside Act 1949. Responsibility for countryside matters was also spread among a number of government departments and, although considerable progress has been made individually in each sphere of activity, the degree of co-ordination has left much to be desired.

Since the 1940s extra powers have been added and the individual agencies have been updated, but the basic administrative structure has remained intact. The Countryside Commission replaced the National Parks Commission in 1968, with a brief that included the whole of the countryside and not just the National Parks and the AsONB. The Town and Country Planning Act 1968 set in train the gradual replacement of Development Plans by Structure and Local Plans, thereby offering land use planners the prospect of their playing a more central strategic role in the development of the landscape. Since the mid-1960s both the Forestry Commission and the Water Authorities have assumed a wider role, so that they now not only provide timber and water respectively, but also actively pursue measures for conserving the landscape and opening their land for recreation. Farming has changed too, but more slowly. Although the threat of food shortages prevalent in the immediate post-war years has receded, the feeling that the UK ought to be as self-sufficient as possible in foodstuffs has persisted. Even membership of the European Community has done little to alter this and agriculture remains an industry apart, governed by different criteria than other parts of the economy.

Despite these changes a feeling of unease about the future of rural planning in the widest sense has persisted (Gilg 1978b). The government has begun to recognise that the continuing expansion of productivity can only be achieved at the expense of other countryside users and the nature and extent of these conflicts has been one of the main themes in this book (Cmnd 6020 1975, Cmnd 7458 1979). It has also been under pressure from the Countryside Commission and some members of the 1974 Sandford

Committee on National Parks to extend planning controls to forestry, although the proposal has been firmly rejected. In 1976 the DOE argued firmly that the way forward lay with improved management techniques, more recreational access and voluntary agreements, rather than any extension of statutory controls (DOE 1976d).

There have also been two major inquiries into the planning system itself, which even though they were not concerned with rural problems specifically, certainly had a bearing on the future of the countryside. The Dobry report concluded that the system was basically sound, but that its operation could be improved (DOE 1975a). The government rejected most of the detailed proposals, but reiterated its view that local authorities should always grant planning permission, unless there were sound and clear cut reasons for refusal (DOE 1976f). By implication this somewhat weakened the hand of authorities wishing to use planning powers as a means of controlling development in the countryside. The 1977 report of the House of Commons Expenditure Committee also suggested improving rather than radically restructuring the system (House of Commons Paper 395-i 1977). In this instance the government took so little heed of the suggestions that the Committee took the almost unprecedented step of issuing a reply to the official response, so as to underline their concern (Cmnd 7056 1978, House of Commons Paper 564 1978). Indeed, it was clear that the government wanted generally to relax planning controls, particularly over the size and design of buildings, but though the proposals were withdrawn in 1978 because of the outcry from both pressure groups and planners themselves about the detrimental effect of the proposals on the appearance of the rural landscape, they were resurrected in 1980.

Other rural land uses, such as water supply, mining, quarrying and defence have also been the subject of official inquiries in the 1970s. In the case of water it was suggested that there had been a general lack of effective central planning (Cmnd 6876 1977). The reports on mining, quarrying and defence all argued for clearer strategies, so that these activities could be better reconciled with each other and with other national policies (Cmnd 5364 1973, DOE 1976c).

The lack of co-ordination is clear, but there has been some attempt, albeit rather half hearted, to introduce general limits on the way in which the rural landscape is managed. Best known of these legislative clauses is Section 11 of the Countryside Act 1968 which stated that, 'Every minister, government department and public body shall have regard to the desirability of conserving the natural beauty and amenity of the countryside.' The response has, however, been rather a token one and additional measures have been suggested by various sources in the subsequent years. The National Parks Policies Review Committee (the Sandford Report) found that the conflict between recreation and conservation in the National Parks had in some cases become irreconcilable and recommended that, when this happened, conservation should take precedence (DOE 1974d). The government agreed and also declared that it would seek ways and means of improving co-ordination between agricultural and environmental policies (DOE 1976d). In particular the conflict between reclamation and landscape conservation was cited as an example, with special reference to Exmoor, where not only two land users are at odds, but also two government departments, the Ministry of Agriculture and the Department of the Environment.

Very similar conclusions were reached by the House of Commons Expenditure Committee's 1976 report on National Parks and the countryside (House of Commons Paper 433 1976). The members were so alarmed at the evidence of unresolved conflicts that they proposed that all the statutory machinery affecting the countryside be streamlined, and that an attempt be made by the government to formulate an overall rural land use strategy. Neither proposal was accepted and further discussion was effectively forestalled by the report being referred to the Countryside Review Committee.

The Countryside Review Committee was set up in 1974 with the following terms of reference, 'To review in relation to the land outside urban areas, the state of the countryside, and the pressures upon it; to examine the effect of existing policies for, or having an impact upon the countryside and the extent to which they are adequate to contain or modify or accept the pressures; Given the existence of other major policy objectives, including the maintenance of agricultural production, to consider whether changes of policy or of practice are necessary to reconcile these objectives where they conflict with the conservation of the countryside, the enhancement of its natural beauty, and its enjoyment by the public; And to make Recommendations' (Countryside Review Committee 1976).

The Committee is chaired and serviced by the Department of the Environment and includes among its members representatives of the Ministry of Agriculture, and, where necessary, representatives from the Countryside Commission, Nature Conservancy Council, Forestry Commission, Sports Councils and Development Commission.

In its first paper in 1976 the Committee concluded that the countryside was facing increased pressure and that although policies have achieved much in resolving this, an early reassessment of priorities was needed. In particular the Committee pointed out that policies are geared to the achievement of individual goals, which often conflict with each other. It argued for a consensus approach, based on the continued multiple use of land and not on the present trend towards exclusive, sectional use. The point was also made that general policies, such as those on taxation, could have important effects on the countryside and be potent forces for change.

In its four subsequent papers on *Rural communities, Leisure and the countryside, Food production in the countryside*, and *Conservation and the countryside heritage* (Countryside Review Committee 1977a, b, 1978, 1979), the Committee has further developed these themes and invited public discussion, especially about how better to co-ordinate policies for genuine multiple land use. For example, in the case of farming the Committee found that, 'We have a situation where for the first time in our history there is an increasing divergence between farming on the one hand and landscape and nature conservation on the other: and the farmer far from being accepted as the guardian of the countryside, is in danger of being regarded as its potential destroyer.' (Countryside Review Committee 1978, p. 13). The establishment of a more accurate inventory of land uses and land use changes was considered essential, so that the present inadequate information base for decision-making can be improved. However, it was also felt that farmers ought to be provided with co-ordinated advice on food production and other aspects of farm life from a single source, so that environmental considerations might be taken into account from the outset, rather than as obstructions at a later stage in the development of a farmer's plans for change.

The latter point was reiterated in the 1978 report of the Advisory Council on Agriculture usually referred to as the Strutt report which stated that, 'It must be recognised that the countryside has many interdependent uses and that purely sectional approaches to the countryside will inevitably lead to conflict' (Advisory Council 1978, p. 12). Strutt's solutions were a wider role for ADAS, so that it included conservation as well as food production, and the creation of local fora to promote consultation amongst all rural land users.

The precise details of a solution are less important than a general recognition of the problem by government. Despite the welter of advice it has received from a wide variety of different sources, there is still a tendency to rest on its past achievements. Undoubtedly it is true that the agricultural decline of the early years of this century has been reversed, and the widespread rural land use change identified in this work is evidence of that fact. Nor has the revival been achieved totally at the expense of the aesthetic qualities of the landscape. Large areas of rural land are now afforded special protection through the planning system and development has been remarkably restrained and confined to existing settlements. Again the evidence presented here has underlined these achievements.

Nevertheless, more needs to be done if all the new demands on the countryside, for recreation space, for raw materials, for water, for defence land and much else, are to continue to be met in the future. The late 1970s have seen the birth of a movement dedicated to producing a new approach to the environment in the 1980s. LAND (Land Use Action for the Next Decade) has an eight-point programme, which basically aims to improve the quality of information about rural areas, to increase productivity and eliminate waste, to promote conservation, and to enhance opportunities for recreation and leisure (Fairweather 1979). Its aims are certainly ambitious, but it is just the kind of co-ordinated approach, needed to weld the rather isolated achievements of the last 30 years into a coherent whole.

The way forward

From the evidence in this book and the studies discussed above it is possible to outline the key issues that will need to be tackled in the 1980s. It has been repeatedly emphasised that a sectional approach will no longer suffice but even so integration of policy making, let alone management, is extremely difficult given the way in which government is organised. We believe that overall guidance is required at national, regional and local level so that conflicting and mutually exclusive decisions may be identified at an early stage and judged according to a single set of established and well-ordered priorities. At present no agency is empowered to issue such advice and it is unlikely that any of the existing agencies would wish, or be allowed to assume such a responsibility. The Countryside Commission is too small and politically weak, MAFF too often identified with the farming lobby, and the DOE too diverse and urban orientated. The best solution might be an inter-departmental committee of senior civil servants, representing all the interested parties, charged with producing an annual white paper, along the lines of the annual roads white paper, which could then be debated in parliament. Within such an overall framework specific suggestions could be made for each land use sector.

In conclusion, so as to give some flavour of what we envisage such a body achieving, we now put forward some tentative suggestions about the way ahead for the most important land users and the major landscapes in the United Kingdom.

Farming. There should be regional variations in price support, grants and subsidies, something that is already done for milk. Production grants should be replaced by landscape maintenance grants in the National Parks and other important habitats, such as the Norfolk Broads and the Somerset Levels. In the non-designated uplands, where production grants probably ought to be retained, an element could be included for landscape management and tree planting. Experimental husbandry farms should be encouraged to try out low energy, self-sufficient farming systems.

Forestry. Grants for hardwood planting on lowland farms should be much increased. Afforestation of the uplands outside the National Parks should continue and actually be increased in Scotland. In the National Parks and similar areas mixed forestry should be encouraged to help reclaim the poorest moorland and eroded hillsides.

Water. Multiple use of water resources ought to be the norm, especially on new reservoirs. More large water recycling schemes ought to be planned, instead of large new reservoirs.

Development. Sporadic development in the countryside should continue to be discouraged and the loophole of agricultural need more rigorously examined. The key settlement policy concept ought to be relaxed. Regional plans, drawn up by consortia of county councils, should replace structure plans and should designate special areas for detailed planning by the relevant individual (groups of) district councils. Conversions of old barns and other redundant buildings should be encouraged.

Recreation.   There should be better provision, with purpose-built facilities provided by local authorities, statutory undertakers and private developers, in the more accessible areas. This is particularly important in the fringe zone between the agricultural lowland and the uplands proper.

Conservation.   As many of the Nature Conservation Review sites as possible should be acquired for the nation, but at the same time there should be inducements available to landowners for retaining and improving as much of the natural habitat as possible.

The variety of landscapes in the United Kingdom inevitably means that no one set of policies devised for individual sectors is going to be adequate; each of the major landscape types will require particular emphases and a different set of priorities:

The urban fringe.   Further development here must be discouraged if the continuing containment of urban Britain is to remain a reality. However, the growing demand for recreation access and facilities cannot be ignored and conveniently swept under the carpet and there is an urgent need to reconcile these demands with genuine agricultural use. The type of landscape created by 'horseyculture' is often very degraded with poor standards of management of both soil and vegetation. One solution may be to pay higher MAFF grants in such areas to offset the additional burdens faced by farmers.

The farmed lowland.   This is the least formally planned of all the country-side landscapes, yet it covers half of the total rural area. A more positive approach to planning and management is much needed to counteract the effects of long-standing planning neglect and farming expansion. County Structure Plans should include landscape management plans with tree planting schemes and measures for the protection of the more valuable landscapes. Government grants ought to have a landscape management element included as a matter of course, and caution should be exercised over village and country town expansion schemes. The growth that is allowed, ought to be concentrated in existing towns and policies should be designed to forestall dormitory settlements, likely to generate large amounts of commuter traffic.

The upland fringe. In England and Wales this landscape forms the border of much of the National Parks and, as such, is a natural buffer zone between the core upland areas and the varied and insistent demands of itinerant visitors and local residents. Their position makes these landscapes both sensitive and crucial and a good deal of attention needs to be paid to them to ensure that policies are formulated and implemented that lead to a diverse environment in which recreation, conservation and agriculture can co-exist. The emphasis should be on variety and flexibility, so as to allow for regional differences in demand and changing circumstances over time. In other words these landscapes should be inherently functional, thus maintaining the longstanding tradition of mixed land use on the fringes of the British uplands.

The uplands. Of all the major landscape types, these are the most fragile, the most contentious and the least susceptible to blanket remedies. In broad terms they comprise three main sub-types:

(a) The rolling uplands of south-west and northern England, southern Wales and southern Scotland;
(b) The more rugged hills of Snowdonia, the Lake District, and north-west Scotland;
(c) The large expanse of the Scottish Highlands.

The rolling uplands are invariably surrounded by extensive farmed lowland and threats to their apparent isolation and the possibility of reclamation causes much over-emotive argument. In reality, however, they are usually made up of a patchwork of permanent moorland, reclaimed farmland and reverted farmland, producing a mixed and very variable landscape. If their future is to be assured, a decision needs to be made as to where the boundary of true moorland lies and the land included within it must then be managed to sustain wildlife, to encourage unenclosed hill farming, and thereby allow free access for recreation. At present it is too early to say whether the new National Park Authorities are sufficiently robust to provide this kind of management framework. If they are to be successful, they need to reverse 2000 years of persistent degradation and to encourage farming practices that will rescue the land from its present impoverished state.

The main value of the rugged hills is their wildness and this needs to be recognised as a key element in any management policies. Nevertheless, selective afforestation and even reservoir construction on a small-scale could be allowed without serious damage being caused. The overriding aim should be to encourage extensive land uses, which do not involve frequent human interference, while at the same time not discouraging access for those seeking solitude as part of their recreation.

The Scottish Highlands are somewhat different. Here the degraded landscape produced by the clearances could be much improved by large-scale afforestation schemes, with little danger of detracting from its intrinsic beauty or its sense of wilderness.

All these suggestions are of course speculative expressions of opinion and ignore many of the more obvious political, social and economic constraints, which determine the future of the rural landscape in practice, but it is the very lack of any consensus about the future of the rural landscape in Britain that gives the greatest cause for concern. There is little argument about the nature of the pressures on the countryside; their precise impact on the landscape is much less certain. This book has tried to marshal a wide range of empirical evidence to throw some light on this whole question; what is needed now is a policy for the future.

Appendix A *The Devon field surveying method*

The field survey had three main aims:

(a) To provide information for the study of development control, in particular design changes and village layouts;
(b) To provide further information on the changing land use structure of rural areas;
(c) To provide an assessment of the impact of planning powers of all kinds on the changing rural landscape.

The survey method involved a traditional ground based land use survey which was conducted at the same time as the survey of the built environment and development control, so that time in the field was put to the fullest use. The land use categories were kept to eleven, so as to speed up the work rate and to follow the categories of land use described by the Coventry–Solihull–Warwickshire landscape evaluation method. The eleven categories each of which were given code numbers for computer analysis of land use totals and change were:

    (0) Orchards,
    (1) Farmland,
    (2) Deciduous woodland,
    (3) Coniferous woodland,
    (4) Developed land including transport,
    (5) Parkland,
    (6) Heathland including rough grazing,
    (7) Water,
    (8) Unused land including land under construction,
    (9) Industrial land,
    (10) Mining land.

Each of these was recorded by coloured crayon on the Ordnance Survey 1:25 000 map, the most suitable scale for broad brush land use surveys not concerned with individual crops. Accordingly, the survey was able to proceed very quickly and in combination with the other surveys.

Although a land use survey can be conducted rapidly and efficiently, the analysis of land use and land use change still presents problems of many kinds. First, one land use type, normally either farmland or moorland dominates, and other uses

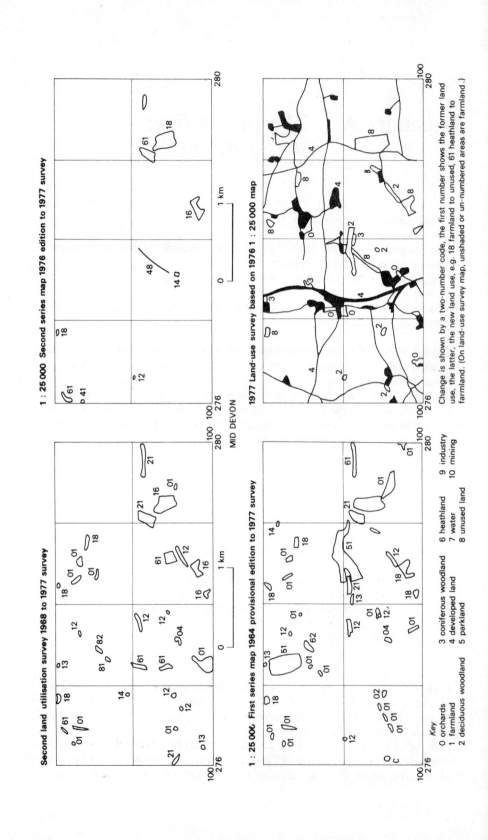

Second land utilisation survey 1968 to 1977 survey

1 : 25 000 Second series map 1976 edition to 1977 survey

MID DEVON

1977 Land-use survey based on 1976 1 : 25 000 map

1 : 25 000 First series map 1964 provisional edition to 1977 survey

Key
0 orchards
1 farmland
2 deciduous woodland
3 coniferous woodland
4 developed land
5 parkland
6 heathland
7 water
8 unused land
9 industry
10 mining

Change is shown by a two-number code, the first number shows the former land use, the latter, the new land use, e.g. 18 farmland to unused, 61 heathland to farmland. (On land-use survey map, unshaded or un-numbered areas are farmland.)

occur sporadically and often in awkward shapes, making small-scale map analysis very difficult. Secondly, land use maps are not easy to analyse quantitatively unless they are first coded and transformed into a digital form by grid square or preferably by grid reference values using a digitiser. Thirdly, land use change cannot be fully assessed because previous surveys have usually been conducted with different definitions.

It is hoped that the method used here has overcome most of these problems. The following procedure was used:

(1) Counting up the area of each land use in each one kilometre square (100 ha) using a 100 one hectare grid overlay.
(2) Recording land use change by comparing land use from the field survey with land use on the 1:25 000 map by (a) using a transparent overlay and (b) by recording the grid square, area and code number for each land use change on punched cards.
(3) Using the punched cards to produce line printer maps showing the distribution of each land use category, and land use change for each of the 110 possible permutations.

This procedure does however have some faults, though on further inspection it was found that these faults could be largely discounted. First, grid overlays are less accurate than digitisers, but for broad brush surveys they are perfectly adequate, especially since each square is self-checked in that it must total to 100 ha. Secondly, land use change may be difficult to assess from a base Ordnance Survey map, since the 1:25 000 map is not primarily a land use map although it has more land use information than any other. However, for the limited categories chosen, no real problem was encountered, and furthermore, other land use surveys were either inconsistent with the definitions used in the authors' survey, or were found to be at variance with the land use patterns mapped by the authors as shown in Figure A.1. In addition, the First Land Use Survey used different scales, and recorded land use at a very low ebb; while the Second Land Use Survey maps did not provide a continuous cover for all the authors' areas for contemporary dates, and many of the areas were only covered by the unpublished 1:10 560/1:10 000 survey sheets held at King's College London. For the sake of convenience, consistency and the level of accuracy required it was concluded that Ordnance Survey maps represented the most cost-effective base. Nonetheless, these maps still have one major disadvantage, namely inconsistent revision dates, a reflection of the chequered history of the 1:25 000 series.

The 1:25 000 series was first launched in the early post-war years as a development of a stop gap wartime series. The 'First Series' (Provisional edition) was drawn from 1:10 560 maps dating from before 1939, and often from the turn of the century, but it was fully revised between 1950 and 1965 using data collected for the

Figure A.1 Land use changes in mid-Devon, using three different base maps. The diagram shows the results of the authors' 1977 survey and changes inferred by comparing this with the 1968 Second Land Utilisation Survey, the First Series Ordnance Survey 1:25 000 map (both of which produce quite different results although they are roughly contemporaneous), and the Second Series Ordnance Survey 1:25 000 map. (Authors' survey and unpublished Second Land Utilisation maps held at the Department of Geography, Kings College, London.)

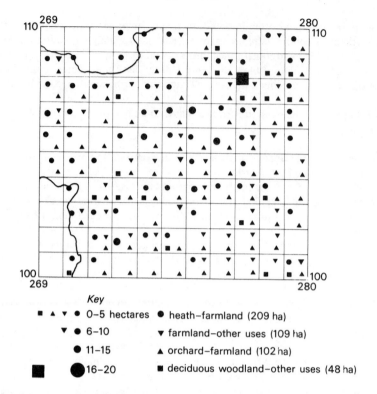

Key

■ ▲ ▼ ● 0–5 hectares   ● heath–farmland (209 ha)

▼ ● 6–10   ▼ farmland–other uses (109 ha)

● 11–15   ▲ orchard–farmland (102 ha)

■ ● 16–20   ■ deciduous woodland–other uses (48 ha)

Figure A.2 Transfer of land use between different categories in the mid-Devon study area. (Authors' survey.)

revision of the 1:63 360 series. Since 1965 the 'First Series' has been gradually replaced with a 'Second Series' (Regular edition) drawn from data collected for the revision of the 1:10 560/1:10 000 series. As Figure A.1 reveals there are considerable differences between the land use changes revealed by comparing 1977 land use with (a) the Provisional edition and (b) the Second Series 1:25 000 maps. In summary, 1:25 000 maps can show up to three periods of land use, although each sheet must be treated on its merits before the following guideline is used:

> Period One: 1900 to 1930s land use; First Series.
> Period Two: 1950s land use; First Series (Revision).
> Period Three: 1970s land use; Second Series.

The main problem with these revision dates is that quite small areas can contain examples of all three periods. In time the Second Series will overcome this and become a very useful base for land use change exercises. Nonetheless, the existing maps did provide useful data for the four areas of Devon surveyed by the authors, particularly when they were analysed in the way shown in Figure A.2. This map and similar ones for the other three study areas, Dartmoor, Exmoor and east Devon, have been used to evaluate the changes discussed in Chapter 4.

In parallel with the land use survey, a survey of planning application sites and all

Table A.1 Form used in authors' dwelling survey.

UNIVERSITY OF EXETER   (FORM A)

DEPARTMENT OF GEOGRAPHY SOCIAL SCIENCE RESEARCH COUNCIL PROJECT
SURVEY OF HOUSE TYPE

SQUARE _____   DATE _____   OPERATOR _____

◄──────── Planning Application Houses ────────►

(1) Pre-1914	*(2)* Interwar	*(3)* Post-war till 1964	*(4)* Planning application	*(5)* Grid ref. + code number	*(6)* State of construction	*(7)* Prominence

Key

Type		Height		Settlement		Walls colour in brackets		Roof		State of construction	
terrace 1(no. of units)		bungalow	1	key settlement	1	brick	1( )	tile	1	built	1
semi-detached	2	2 storey	2	non-key settlement		stone	2( )	slate	2	under construction	2
detached	3	3 or more	3	village	2	render	3( )	tin	3	foundations	3
caravan	4	*Garage*				cob	4( )	thatch	4	no start	4
farmhouse	5	none	1	hamlet	3	mixture	5( )	asbestos	5	*Prominence*	
		1	2	group	4	stone-				very	1
		2 or more	3	isolated	5	faced	6( )			fairly	2
						wood	7( )			not at all	3
						tile	8( )				
						iron	9( )				
						asbestos	10( )				

existing dwellings was made, using the form shown in Table A.1. On this form, a detached bungalow with a garage, located in a key settlement, with red brick walls and a tile roof would be coded as 31211(R)1. In addition each area was surveyed and each dwelling classified according to the four ages shown in Table A.1.

All of the data was thus recorded and analysed by computer using the grid square mapping technique. Any possible permutation of land use change, building design and layout can thus be tested at will, and the authors have a wealth of detailed data that space has precluded from presenting in this text, although the computer has allowed the main themes to be summarised.

Appendix B   *A computer based information system for the retrieval and analysis of planning applications*

The system was initially derived to provide information for a one year pilot project, financed by the Social Science Research Council, which studied the effect of planning decisions on the landscape of east Devon. The period of study was from 1967–72 and the analysis was undertaken in the twelve months from August 1973. The system was extended for the larger project, covering Dartmoor, Exmoor and mid-Devon between 1975 and 1979. Although this appendix deals with east Devon in particular, the procedure described is identical to that used in the other study areas.

The system is based on a file of códed record cards compiled by the pre-1974 Devon County Council, but it could easily be adapted to fit other local authority records.

Preliminary sorting

The Devon records are punched onto data cards with the following layout:

EM/15066 UNIGATE LIMITED,
EM/15066 ABB. ERCTN OF DSTBTN WHSE INCLDNG OLD STRS, OFFICES
          & VHCLE WRKSHOP,
EM/15066 KESTREL WAY, SOWTON INDUSTRIAL ESTATE, SOWTON,
EM/15066 20 10 71   18   1 72,
15066   2   5   10   39   56   84   88   92   96.

Thus the first data card deals with the name of the applicant, the second with the nature of the application (in many cases this is preceded by ABB to denote the abbreviation necessary to condense the application details into the 80 spaces available on a computer card). The third with the location of the application, the fourth with the dates on which the applications were received and the date of decision, and the fifth with the codes punched on the Devon record cards. These codes are numbered in ascending order from the top left hand corner in a clockwise direction (but ignoring the 4 corner spaces, see Fig. B.1). These cards are then read onto a master magnetic tape. Processing and analysis can now begin.

Initially, the applications are recorded by the date of decision. Secondly, each application is given a grid reference, sadly missing from these particular Devon records, so that spatial analysis techniques can be employed (and so that the data is directly comparable with other data collected for the information system). The data is thus also comparable with census data collected by grid squares.

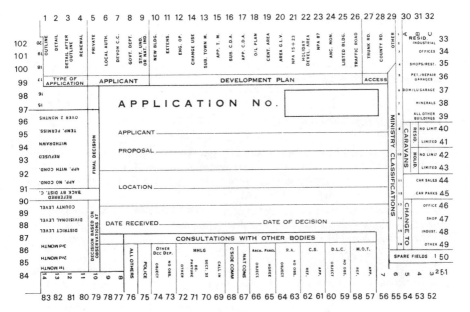

Figure B.1 Devon County Council planning application coding system.

The acquisition of the grid references was a laborious process, cutting substantially into the time saved by the rapid coding up of the Devon cards (20 000 cards in 3 weeks). The method of acquisition was threefold. First the 1:2500 maps on which the planning applications were recorded in the files of the St Thomas Rural District Council were scanned for those EM numbers which appeared to fall within the range of our study (1967–72). Each application number was punched up on data cards with its appropriate grid reference or grid references. Applications which have two or more grid references are common, the most typical case being applications by the South-Western Electricity Board (SWEB). When the two files were matched up by program GG 1381 about 20% of the applications were found to lack a grid reference. Subsequent rescanning of the 1:25 000 map reduced this figure to about 10%. To find values for the final 10%, all the planning applications recorded on the 1:2500 maps were collected (providing a useful data base for comparing the rate of applications over the whole period) and processed (GG 1384). The few remaining records not obtained from this process were obtained by direct contact with the planning officers concerned, the local offices of SWEB and by local enquiries at, for example, post offices. The final insertion of these records and tidying up of other errors is achieved by the first major program of the sequence GG 1380. This initial ordering of the data involved a time span of 12 months, but in real work time about 4 months full-time work for all the operators involved. If a working time of 8 hours a day is computed with the number of working days in 4 months (84 days), this gives a total of 672 hours. This produces a rate of 5·3 applications processed per working hour, or 42·8 applications per day. This compares with a provisional estimate of 20 applications per day experienced by David Gregory and his research team when working directly from unprocessed local authority applications in the West Midlands. The figures are not strictly comparable since our figures do not include the time taken by Devon County Council to process the record cards and David

Gregory's do not include the time taken to punch up the coding forms. Both figures are of course vastly greater than would be experienced by a local authority processing the applications as they were being dealt with. A realistic time input would be 10 minutes per application, to include coding up, punching, verifying and insertion of the data cards into the main data set held on magnetic tape.

The magnetic tape operating system

The information system involves the use of 8 magnetic tapes, 2 long tapes containing all the details involved and 6 short tapes containing data broken down into type or area. The following is a summary of the process involved, and should be read with reference to the flow diagram (Fig. B.2).

The diagram begins at the stage when program GG 1381 has matched up nearly all the applications with their appropriate grid references.

The first program of the series GG 1380 performs two functions:

(i) It inserts all remaining grid references, and updates and corrects the existing file;
(ii) It increases the amount of data describing the application so that the record for each application reads as follows:

Record 1 J and NCOUNT number. J refers to the total number of discrete applications processed by the file at any one time, thus the first application is given a value of 1. NCOUNT refers to the number of grid references used, so for example if the third application covered two grid references the J value would remain at 3, but there would be two NCOUNT values 4 and 5, and two copies of the remaining records described below:

Record 2 EM/---- name of applicant,
Record 3 EM/---- nature of application,
Record 4 EM/---- location of application,
Record 5 EM/---- dates when application received and decided,
Record 6 EM value, up to 18 CODE values, two grid references CODE (19), CODE (20) and a value for the number of units involved in housing or garage applications CODE (21).

The next program in the sequence is GG 1382 which updates and reorders the two major tape files held on EXT 121 and EXT 129, in a simpler manner than GG 1380. GG 1382 is basically a copying program that inserts certain updated values, and reorders the J and NCOUNT values where the updating sequences of GG 1380 and GG 1382 have altered them. The format and content of the tapes is otherwise not altered by this program. Figure B.2 shows that by using two tapes and two magnetic tape drives the process of editing and updating the tape files to a final correct session took about 2 working weeks, or in run times about 8 runs of the program. The first run (4 October 1974) was a simple run transferring the data sent by 1380 onto EXT 121 back onto EXT 129, in case the data on EXT 121 was destroyed. The job of successfully running the updating version of GG 1382 took from then till 21 October 1974. The provision of a second copy on EXT 129 only took a further 2 days. At this stage EXT 129 and EXT 121 are identical and from this stage *only* EXT 129 is used,

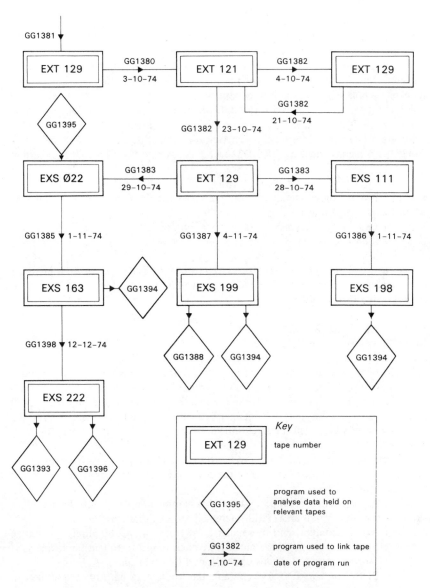

Figure B.2 Flow diagram of magnetic tape operating system.

EXT 121 being kept as a spare copy in case EXT 129 became corrupted or otherwise unusable. Two master files are vitally necessary in this sort of complex operation. The master tape EXT 129 however contains 3600 NCOUNT values, and takes a considerable time to read, so that the next stage of the system divides the master tape into area and content files.

Program GG 1383 divides the master file into two area types. The Non-Field Area (NFA) of St Thomas RDC and the Field Area (FA) of St Thomas RDC, the area covered in detail by the other SSRC work on landscape evaluation and surveys on dwelling stock. The contents of these files are identical to EXT 129 and EXT 121 with one exception. On EXS 111 (NFA tape) record 1 is altered to J, NOUT, NCOUNT, J and NCOUNT remaining the same as before, with NOUT performing the same function as NCOUNT for the smaller file. NOUT thus records the total number of grid references and thus application records on this file. On EXS Ø22 (NFA tape) NOUT is replaced with NIN, which performs the same function.

Both EXS Ø22 and EXS 111 are now divided by content.

Program GG 1385 performs this function for EXS Ø22, and transfers and transforms the 6 records held on EXS Ø22 to 2 records held on EXS 163, these are as follows:

Record 1 EM/---- dates when application received and decided, J, NIN, NCOUNT, Record 2 as record 6 in EXT 129,
Program GG 1386 performs the same function for EXS 111 and EXS 198 and produces the same record.

The same 2 record type format is produced by program GG 1387 which transforms and transfers the 6 records held on EXT 129 onto EXS 199, so that EXS 199 contains the following two records:

Record 1 EM/---- J, NCOUNT,
Record 2 as record 6 in EXT 129.

Finally because some of the values held on Record 1 on EXS 163 are superfluous to analysis, program GG 1398 was used to transfer only Record 2 on EXS 163 onto EXS 222. The long time delay is due to the fact that a month earlier tape EXS 2ØØ had been used for this purpose but had become corrupted thus necessitating the creation of a new tape. This illustrates the need for 2 tapes if at all possible for every data set.

The operating system has thus produced a powerful data bank for subsequent analysis, by the programs depicted in the diamonds. The data has been organised into three basic subsets.

Subset A: By smaller area, so that analysis can be compared with detailed field survey within the overall area. Grid references are essential here.

Subset AB: By content, and by smaller area. In this stage the purely qualitative data is left out leaving only those numerical values useful for quantitative analysis.

Lists of applications and details can be found by using subset A.

Subset B: By content, EXS 199 contains the whole area but only the numerical values.

Programs used for analysis

The following programs were used to analyse the information held on tape system, these programs are those listed inside the diamonds in Figure B.2.

(i) GG 1388 is an indexing program which ranks all the EM numbers into ascending order and lists the appropriate J number (ranked by date of decision) so that the location of any application on say either tape EXT 129 or EXS Ø22 can be found very easily. Where an EM number occurs twice the J number is preceded by 90 and reference is made to a separate list showing the J numbers for those EM numbers that are duplicated on the master tape.

(ii) Further qualitative analysis is achieved by GG 1395 which searches within the radius of a circle or a rectangle for all applications or for a certain type of application with a particular set of CODE values. For each application all 6 records are printed out. The circle can be of any radii and can have any central point defined by the National Grid. The square is similarly defined by the National Grid. Quantitative analysis is achieved by programs GG 1396, GG 1394 and GG 1393.

(iii) The next program GG 1396 takes each of the 100 CODE values in turn and lists the EM numbers that contain such CODE values for each 1 kilometre square of the National Grid. If further details are required use can be made of the index GG 1388 and from this the EM number can be traced on say EXT 129. At the end of each listing a map is produced showing the distribution of CODE values by grid square. Two versions of this program exist, the first ignores the weighting value contained in CODE (21) which specifies the number of houses or garages or caravans applied for, the second takes this into account and multiplies the application accordingly. However, in both versions where an application is given two or more grid references and is thus repeated, only the first application is counted. This form of analysis can throw a good deal of light on the 'accuracy' of planning application records in terms of recording real demand, for published totals do not weight applications by size or importance.

(iv) The next program GG 1394 extends this idea further by taking samples to see how accurately these reflect the total, and by looking at the same data in 4 different ways.

NTEST/CODE (21)	EXS 199	EXS 163	EXS 198
NO NO	10	11	12
YES NO	3	2	1
NO YES	9	8	7
YES YES	4	5	6

The data refer to the run numbers of program GG 1394 from 1 to 12. It can be seen that Run 1 contained NTEST but did not contain CODE (21), as did Run 2. NTEST is the name for the subprogram check which eliminates the second and further records of those EM numbers that are duplicated. Thus an electricity board application for a line of pylons may have 6 sets of grid references, and be repeated 6 times. Arguably it is 6 different applications involving 6 widely spaced locations, but also arguably it is one linked application and should only be counted once. Accordingly GG 1394 allows both assumptions to be tested by either including the NTEST check or deleting it. The CODE (21) operation has already been explained for GG 1396. The results of these runs show very wide differences between the 4 possible permutations, emphasising again the need for such accounting procedures to become a part of local authority record systems.

GG 1394 produces two sets of results, a straight list of the number of applications containing each of the 100 CODE values, thus in run 12 CODE (1) occurred 676 times. The main set of results is, however, a matrix of CODE values so that the number of times CODE (44) was paired with CODE (72) for example can be assessed. Because Devon County Council changed the CODE system halfway through the period of our study this necessitates the output of two blocks of results. This matrix demonstrates the power of this sort of information processing system.

(v) The final program GG 1393 provides the finer geographical mesh that GG 1396 lacks being based on the kilometre square. GG 1393 produces graph plotter maps of applications based on their CODE value, so for example a map of housing applications accurate to 100 m in location may be produced showing with different colour or different size crosses the distribution of those applications granted, granted conditionally or refused. At the same time a list is produced with the details of all 6 records for each application described on the graph plotter map.

Hardware

All these programs are at present run by using cards and magnetic tape input. Output is by line printer so a permanent record of the results of each program is kept, however even for a study involving only 3000 applications the paper for all the operations described here takes up the equivalent of 20 line printer files. The next step in the development of the system is to allow access to the programs by teletype terminal so that one-off results can be printed out for say programs GG 1395 and GG 1393, in the latter case using a video screen. The system needs as a minimum for its development the following equipment:

(a) Card reader,
(b) Line printer,
(c) Graph plotter,
(d) Two magnetic tape drivers (discs would be an alternative),
(e) Computer with at least 128k of core store.

Bibliography

Advisory Council for Agriculture and Horticulture in England and Wales 1978. *Agriculture and the countryside: the Strutt report*. London: MAFF.

Aitken, R. 1977. *Second homes in Scotland*. Dartington: Dartington Amenity Research Trust.

Alexander, K. J. W. 1978. The Highlands and Islands Development Board. In *The future of upland Britain*, R. B. Tranter (ed.). Reading: Centre for Agricultural Strategy.

Ambrose, P. 1974. *The quiet revolution: Social change in a Sussex village*. London: Chatto & Windus.

Anderson, K. 1975. An agricultural classification of England and Wales. *Tijds. Econ. Sociale Geog.* **66,** 148–58.

Anderson, M. A. 1977. A comparison of figures for the land use structure of England and Wales in the 1960s. *Area* **9,** 43–5.

Anon 1978. Moorland conversion. *Farmer's Weekly* July 7, 56.

Ashby, A. W. 1939. The effects of urban growth on the countryside. *Sociol. Rev.,* 345–69.

Ashby, A. W. 1978. Britain's food manufacturing industry. *J. Agric. Econ.* **29,** 213–24.

Ashby, P., G. Birch and M. Haslett 1975. Second homes in North Wales. *Town Planning Review* **46,** 323–33.

Aves, M. 1976. Enforcing Section 52 agreements. *J. Plann. Environ. Law,* 216–23.

Baird, W. and J. Tarrant 1972. Vanishing hedgerows. *Geog. Magazine* **44,** 545–51.

Barrett, E. and L. Curtis (eds) 1974. *Environmental remote sensing*. London: Edward Arnold.

Best, R. H. 1976a. The changing land use structure of Britain. *Town and Country Planning* **44,** 171–6.

Best, R. H. 1976b. The extent and growth of urban land. *The Planner* **62,** 8–11.

Best, R. H. 1978. Agricultural land loss: myth or reality? *The Planner* **63,** 15–16.

Best, R. H. and A. Champion 1970. Regional conversions of agricultural land to urban use in England and Wales 1945–1967. *Trans, Inst. Br. Geogrs* **49,** 15–31.

Best, R. H. and A. Rogers 1976. *The urban countryside: the land-use structure of small towns in England and Wales*. London: Faber and Faber.

Bielckus, C. L., A. W. Rogers and G. P. Wibberley 1972. *Second homes in England and Wales*. Ashford: Rural School of Economics and Related Studies.

Blacksell, M. 1978. Places for leisure: pressure on the Broads. *Geog. Magazine* **L1,** 88.

Blacksell, M. 1979. Landscape protection and development control: an appraisal of planning in rural areas of England and Wales. *Geoforum* **10,** 267–74.

Blacksell, M. and A. W. Gilg 1974. *The influence of planning decisions on the rural landscape of south east Devon* HR 2537/2. London: SSRC.

Blacksell, M. and A. W. Gilg 1977. Planning control in an Area of Outstanding Natural Beauty. *Social and Economic Administration* **11,** 206–15.

Blunden, J. 1975. *The mineral resources of Britain*. London: Hutchinson.

Boddington, M. A. B. 1978. *The classification of agricultural land in England and Wales: a critique*. Oxford: Rural Planning Services.

Bowman, J. C., C. J. Robbins and C. J. Doyle 1976. The case for an agricultural strategy for the United Kingdom. *R. Agric. Soc. J.* **137**, 27–33.

Bracey, H. 1952. *Social provision in rural Wiltshire*. London: Methuen.

Brett, L. 1965. *Landscape in distress*. London: Architectural Press.

British Road Federation 1977. *Basic road statistics*. London: The Federation.

Britton, D. K. and B. Hill 1975. *Size and efficiency in farming*. London: Saxon House.

Budowski, G. 1976. Tourism and environmental conservation. *Environmental Conservation* **3**, 27–31.

Cabinet Office 1976. *Future world trends*. London: HMSO.

Caborn, J. M. 1971. The agronomic and biologic significance of hedgerows. *Outlook on Agriculture* **6**, 279–84.

Campbell, J. 1975. Can private forestry survive high capital taxation? *Q. J. For.* **69**, 195–201.

Capstick, M. 1978. Economic, social and political structures in the uplands of Cumbria. In *The future of upland Britain*, R. B. Tranter (ed.). Reading: Centre for Agricultural Strategy.

Carney, J. and R. Hudson 1978. The Scottish Development Agency. *Town and Country Planning* **46**, 507–10.

Carney, J. and R. Hudson 1979a. The Welsh Development Agency. *Town and Country Planning* **48**, 15–16.

Carney, J. and R. Hudson 1979b. The European regional development fund. *Town and Country Planning* **48**, 125–6.

Carter, C. 1973. Agricultural land drainage. *Power farming* **51**.

Central Office of Information 1974. *Regional development in Britain*. London: HMSO.

Central Office of Information 1975. *Local government in Britain*. London: HMSO.

Central Office of Information 1977. *Agriculture in Britain*. London: HMSO.

Central Statistical Office 1977. *Social trends*. London: HMSO.

Champion, A. G. 1976. Evolving patterns of population distribution in England and Wales 1951–71. *Trans. Inst. Br. Geogrs.* **1**, 401–20.

Cherry, G. 1976. *Rural planning problems*. London: Leonard Hill.

Church, B. 1968. A type of farming map based on agricultural census data. *Outlook on Agriculture* **5**, 191–6.

Civil Service Department 1976. *A directory of paid appointments made by Ministers*. London: HMSO.

Cloke, P. 1977a. An index of rurality for England and Wales. *Regional Studies* **11**, 31–46.

Cloke, P. 1977b. In defence of key settlement policies. *The Village* **17**, 19–31.

Cloke, P. 1979. *Key settlements in rural areas*. London: Methuen.

Coleman, A. 1961. The second land use survey. *Geog. J.* **127**, 168–86.

Coleman, A. 1969. A geographical model for land use analysis. *Geography* **54**, 43–55.

Coleman, A. 1976. Is planning really necessary? *Geog. J.* **142**, 411–37.

Coleman, A. 1978. Last bid for land-use sanity. *Geog. Magazine* **50**, 820–4.

Coleman, A., J. Isbell and G. Sinclair 1974. The comparative statics approach to British land use trends. *Cartographic J.* **11**, 34–41.

Cmd 6153 1940. *Report of the Royal Commission on the distribution of industrial population*. London: HMSO.

Cmd 6378 1942. *Report of the committee on land utilisation in rural areas*. London: HMSO.

Cmd 6628 1945. *National parks in England and Wales.* London: HMSO.

Cmd 6631 1945. *National parks for Scotland.* London: HMSO.

Cmd 7121 1947. *National parks (England and Wales).* London: HMSO.

Cmd 7814 1947. *Final report by the Scottish national parks committee and the Scottish wildlife conservation committee.* London: HMSO.

Cmnd 462 1958. *Report of the Royal Commission on common land.* London: HMSO.

Cmnd 5364 1973. *Report of the Defence Lands Committee.* London: HMSO.

Cmnd 5714 1974. *Statement on the report of the Defence Lands Committee 1971–73.* London: HMSO.

Cmnd 6020 1975. *Food from our own resources.* London: HMSO.

Cmnd 6876 1977. *The water industry: The next steps.* London: HMSO.

Cmnd 7056 1978. *Planning procedures.* London: HMSO.

Cmnd 7058 1978. *Annual review of agriculture.* London: HMSO.

Cmnd 7458 1979. *Farming and the nation.* London: HMSO.

Cmnd 7599 1979. *Report of the Committee of Inquiry into the acquisition and occupancy of agricultural land.* London: HMSO.

Coppock, J. T. 1960. The parish as a geographical/statistical unit. *Tijds. Econ. Sociale Geog.* **51**, 317–26.

Coppock, J. T. 1971. *An agricultural geography of Great Britain.* London: Bell.

Coppock, J. T. 1974. *An agricultural atlas of England and Wales.* London: Faber.

Coppock, J. T. (ed.) 1977. *Second homes: curse or blessing?* Oxford: Pergamon.

Coppock, J. T. 1979. Rural land management in Scotland. *Town and Country Planning* **48**, 47–8.

Coppock, J. and L. Gebbett 1978. *Land use and town and country planning: review of United Kingdom statistical sources.* Oxford: Pergamon.

Cornwall County Council 1976. *Structure plan topic report: environment.* Truro: the Council.

Countryside Commission 1973. *Study of informal recreation in south-east England.* Cheltenham: The Commission.

Countryside Commission 1974. *New agricultural landscapes, CCP76.* Cheltenham: The Commission.

Countryside Commission 1975. *Grants for amenity tree planting, CCP91 and CCP103.* Cheltenham: The Commission.

Countryside Commission 1976. *The Lake District Upland Management Experiment, CCP93.* Cheltenham: The Commission.

Countryside Commission 1977. *New agricultural landscapes: issues, objectives and action, CCP102.* Cheltenham: The Commission.

Countryside Commission 1978a. *Tenth report of the Countryside Commission.* Cheltenham: HMSO.

Countryside Commission 1978b. *Areas of Outstanding Natural Beauty: a discussion paper, CCP116.* Cheltenham: The Commission.

Countryside Commission 1979a. *The Broads,* CCP119. Cheltenham: The Commission.

Countryside Commission 1979b. *Woodland survey of Gwent. A pilot study by DART.* Cheltenham: The Commission.

Countryside Commission for Scotland 1974. *A park system for Scotland.* Perth: The Commission.

Countryside Commission for Scotland 1978. *Scotland's scenic heritage.* Perth: The Commission.

Countryside Review Committee 1976. *The countryside: problems and policies: a discussion paper.* London: HMSO.

Countryside Review Committee 1977a. *Leisure and the countryside.* London: HMSO.

Countryside Review Committee 1977b. *Rural communities.* London: HMSO.

Countryside Review Committee 1978. *Food production in the countryside.* London: HMSO.

Countryside Review Committee 1979. *Conservation and the countryside heritage.* London: HMSO

Cresswell, P. 1975. The South-West plan. *Planning,* 7 November.

Cripps, J. 1980. The Countryside Commission: its first decade. *Countryside Planning Yearbook* **1,** 38–48.

Dahlstein, D. 1976. The third forest. *Environment* **18,** 35–42.

Davidson, J., A. Hall, P. Webster, P. Berry and M. Fitton 1976. The urban fringe. *Countryside Recreation Review* **1,** 2–34.

Davidson, J. and R. Lloyd (eds) 1978. *Conservation and agriculture.* Chichester: John Wiley.

Davies, E. 1976. *The Dartmoor and Exmoor National Parks: changes in farming structure 1952–72.* Exeter: Agricultural Economics Unit.

Davies, E. 1977. *Aspects of land use in the Exmoor National Park Critical Amenity Area.* Exeter: Agricultural Economics Unit.

Davies, S. 1978. Eighth report from the Commons Expenditure Committee: an opportunity for self appraisal? *The Planner* **64,** 146–7.

Denbighshire County Council 1974. *Rural housing in Denbighshire.* Ruthin: the Council.

Denman, D. R. 1977. Who owns Britain? *Countryman* Winter, 23–32.

Dennier, D. A. 1978. National park plans: a review article. *Town Planning Review* **49,** 175–184.

Dennier, D. A. 1980. National Park Plans. *Countryside Planning Yearbook,* 49–66.

Development Commission 1977. *Annual report.* London: HMSO.

Devon County Council 1955. *Dartmoor: building in the National Park.* London: Architectural Press.

Devon County Council 1964. *Development plan: first review.* Exeter: the Council.

Devon County Council 1969. *Development plan: second review.* Exeter: the Council.

Devon County Council 1977. *Report of survey.* Exeter: the Council.

Devon County Council 1979. *County structure plan.* Exeter: the Council.

Dickinson, G. and M. Shaw 1978. The collection of national land use statistics. *Environment and Planning A* **10,** 295–303.

Dickinson, R. 1932. The distribution and function of the smaller urban settlements of East Anglia. *Geography* **17,** 19–31.

DOE (Department of the Environment) 1971a. *General information system for planning.* London: HMSO.

DOE 1971b. *Long-term population distribution in Great Britain.* London: HMSO.

DOE 1971c. *Town and Country Planning Acts 1962–1968.* Circular 80/71. London: HMSO.

DOE 1973. *Structure plans: the examination in public.* London: HMSO.

DOE 1974a. *Development control policy notes.* London: the Department.

DOE 1974b. *Local government in England and Wales: a guide to the new system.* London: HMSO.

DOE 1974c. *Nature conservation.* Circular 161/74. London: HMSO.

DOE 1974d. *Report of the National Parks Policies Review Committee.* London: HMSO.

DOE 1974e. *Statistics of land-use change.* Circular 71/74. London: HMSO.

DOE 1975a. *Review of development control: final report by George Dobry.* London: HMSO.

DOE 1975b. *Review of the development control system.* Circular 113/75. London: HMSO.

DOE 1976a. *Development involving agricultural land.* Circular 75/76. London: HMSO.

DOE 1976b. *National land use classification.* London: HMSO.

DOE 1976c. *Planning control over mineral working.* London: HMSO.

DOE 1976d. *Report of the National Parks Policies Review Committee.* Circular 4/76. London: HMSO.

DOE 1976e. *Spatial retrieval for point referenced data.* London: HMSO.

DOE 1976f. *The Dobry report: action by local authorities.* Circular 9/76. London: HMSO.

DOE 1977a. *A study of Exmoor.* London: HMSO. See Porchester.

DOE 1977b. *General Development Order.* SI 289/77. London: HMSO.

DOE 1977c. *Memorandum on structure and local plans.* Circular 55/77. London: HMSO.

DOE 1977d. *Nature conservation and planning.* Circular 108/77. London: HMSO.

DOE 1978a. *Consultative document on common land.* London: the Department.

DOE 1978b. *Developed areas 1969: a survey of England and Wales from air photography.* London: the Department.

DOE 1978c. *Trees and forestry.* Circular 36/78. London: HMSO.

DOE 1980. *Development control – policy and practice.* Draft circular. London: the Department.

Dower, M. 1977. Planning aspects of second homes. In J. T. Coppock, *op. cit.,* 155–64.

Downing, P. and M. Dower 1974. *Second homes in England and Wales.* Cheltenham: the Countryside Commission.

Draper, D. 1977. *The creation of the DOE.* London: HMSO.

Drewett, R. 1969. A stochastic model of the land conversion process. *Regional Studies* **4**, 269–80.

Economist Intelligence Unit 1978. *Land availability. A study of land with residential planning permission.* London: DOE.

Edwards, A. J. 1969. Field size and machine efficiency. In *Hedges and Hedgerow Trees*, M. D. Hooper and M. W. Holdgate (eds). London: Nature Conservancy Council.

Essex County Council 1976. *Essex exercise: farming, wildlife and planning.* Chelmsford: The Council.

European Communities 1978. *Importance and functioning of the European Agricultural Guidance and Guarantee Fund.* London: HMSO.

Evans, A. F. 1969. The impact of taxation on agriculture. *J. Agric. Econ.* **20**, 217–28.

Eversley, D. 1973. *The planner and society.* London: Faber.

Eversley, D. 1975. Regional devolution and environmental planning. In *Regional devolution and social policy*, E. Craven (ed.), 35–38. London: Macmillan.

Exmoor National Park Committee 1977. *Exmoor National Park plan: supplement.* Dulverton: The Committee.

Exmoor Society 1966. *Can Exmoor survive?* Minehead: The Society.

Expenditure Committee: Eighth Report of 1977. *Planning procedures.* HC 395–I, –II and –III. HC 35 (1976–77) and HC 466 (1975–76) London: HMSO.

Experimental Cartography Unit 1978. *Land use mapping by local authorities in Britain.* London: Architectural Press.

Fairweather, L. 1979. Land of our fathers: fit for our children. *Architects Journal* **169**, 1307–1314.

Farquarson, J. K. 1978. The future of the Highlands and Islands Development Board in the encouragement of rural based industries. In *The future of upland Britain*, R. B. Tranter (ed.). Reading: Centre for Agricultural Strategy.

Feist, M. J. 1979. *A study of management agreements.* Cheltenham: The Countryside Commission.

Feist, M. J., P. M. K. Leat and G. P. Wibberley 1976. *A study of the Hartsop Valley.* Cheltenham: the Commission.

Finer, S. E. 1974. *The study of interest groups.* London: Macmillan.

Fordham, R. 1975. Urban land use changes in the United Kingdom. *Urban Studies* **12**, 71–84.

Forestry Commission 1971. *Forest management for conservation, landscaping access and sport.* Edinburgh: The Commission.

Forestry Commission 1972. *The new forests of Dartmoor.* London: HMSO.

Forestry Commission 1974a. *British forestry.* London: HMSO.

Forestry Commission 1974b. *Forest products in the United Kingdom economy.* London: HMSO.

Forestry Commission 1975. *Explore the New Forest.* London: HMSO.

Forestry Commission 1977a. *Advice for woodland owners: Basis II and Basis III Dedication schemes: New small woods scheme.* Edinburgh: The Commission.

Forestry Commission 1977b. *Annual report 1976–77.* London: HMSO.

Fudge, C. 1976. Local plans, structure plans and policy planning. *The Planner* **62**, 174–6.

Gasson, R. 1969. Occupational immobility of small farmers. *J. Agric. Econ.* **20**, 279–88.

Gasson, R. 1973. Goals and values of farmers. *J. Agric. Econ.* **24**, 521–37.

Gasson, R. 1975. *Provision of tied cottages.* Cambridge: University of Cambridge Department of Land Economy.

Geddes, P. 1949. *Cities in evolution.* London: Williams and Norgate.

Gibbs, R. and A. Harrison 1974. *Land ownership by public and semi-public bodies in Great Britain.* Reading: Department of Agricultural Economics.

Gibbs, R. and M. Whitby 1975. *Local authority expenditure on access land.* Newcastle-Upon-Tyne: Agricultural Adjustment Unit.

Gilg, A. W. 1975. Regional planning in the south-west. *Social and Economic Administration*, 220–5.

Gilg, A. W. 1976. Rural employment. In *Rural planning problems*, G. Cherry (ed.), 125–76. London: Leonard Hill.

Gilg, A. W. 1978a. *Countryside planning.* Newton Abbott: David and Charles.

Gilg, A. W. 1978b. Needed a new Scott Report. *Town Planning Review* **5**, 353–71.

Gilg, A. W. (ed.) *Policies for landscapes under pressure.* Bury St. Edmunds: Northgate.

Gilg, A. W. 1980. Annual review of the countryside. *Countryside Planning Yearbook* **1**, 9–36.

Glasson, J. 1977. Regional aid from the EEC. *Town and Country Planning* **45**, 167–70.

Goodall, B. 1973. The composition of forest landscapes. *Landscape Res.* **1**, 6–10.

Goodier, R. (ed.) 1971. *The application of aerial photography to the work of the nature conservancy.* London: Nature Conservancy.

Grant, M. 1977. The European regional development fund. *J. Environ. Plann. Law*, 232–4.

Green, F. 1976. Recent changes in land use and treatment. *Geog. J.* **142**, 12–26.

Green, R. 1971. *Country planning: the future of rural regions.* Manchester: University Press.

Green, R. 1980. Planning the rural sub-regions. *Countryside Planning Yearbook* **1**, 67–85.

Gregory, D. G. 1970. *Green belts and development control.* Birmingham: Centre for Urban and Regional Studies.

Grieve, R. 1973. Scotland: Highland experience of regional government. *Town and Country Planning* **41**, 172–6.

Grigg, D. B. 1976. The world's agricultural labour force 1800–1970. *Geography* **61**, 194–202.

Hall, P., R. Thomas, H. Gracey and R. Drewett 1973. *The containment of urban England.* London: George Allen & Unwin.

Hamersley, C. 1974. Timber and the crisis of scarce resources. *Town and Country Planning* **42**, 307–10.

Harrison, M. L. 1972. Development control; the influence of political, legal and ideological factors. *Town Planning Review* **43**, 254–74.

Harrison, A., R. B. Tranter and R. S. Gibbs 1977. *Landownership by public and semi-public institutions in the UK.* Reading: Centre for Agricultural Strategy.

Hart, P. 1977. Countryside implications of the decisions of rural landowners. *Area* **9**, 45–8.

Health, Ministry of 1939. *Report on the preservation of the countryside.* London: HMSO.

Health, Ministry of 1944. *Rural housing.* London: HMSO.

Heap, D. 1976. *An outline of planning law.* Andover: Sweet and Maxwell.

Hebbert, M. and I. Gault 1978. *Green belt issues in local plan preparation.* Oxford: The Polytechnic.

Helme, W. H. 1975. The agricultural development and advisory service. *J. Agric. Econ.* **26**, 53–60.

Hereford and Worcester County Council 1976. *New agricultural landscapes.* Worcester: The Council.

HM Treasury 1972. *Forestry in Great Britain: an interdepartmental cost/benefit study.* London: HMSO.

Hookway, R. J. S. 1967. *The management of Britain's rural land.* London: Town and Country Planning Association.

Hookway, R. J. S. 1978. National park plans. *The Planner* **64**, 20–2.

Hookway, R. J. S. and J. Hartley 1968. Resource planning. *Estates Gazette*, 13 July.

House of Commons Paper 511 1972. *Report of the Select Committee on Scottish Affairs.* London: HMSO.

House of Commons Paper 433 1976. *Sixth report from the Expenditure Committee of the House of Commons: National parks and the countryside.* London: HMSO.

House of Commons Paper 256 1977. *Observations by the Secretaries of State for the Environment, for Scotland and for Wales on the sixth report of the Expenditure Committee on national parks and the countryside.* London: HMSO.

House of Commons Paper 395-i 1977. *Eighth report of the Expenditure Committee of the House of Commons: planning procedures.* London: HMSO.

House of Commons Paper 564 1978. *Eleventh report from the Expenditure Committee of the House of Commons.* London: HMSO.

House of Lords Papers 97 and 97-i 1977. *Report 23: EEC farm prices 1977–78.* London: HMSO.

Housing and Local Government, Ministry of 1962. *The Green Belts.* London: HMSO.

Housing and Local Government, Ministry of 1965. *Planning advisory group: the future of development plans.* London: HMSO.

Housing and Local Government, Ministry of 1967. *Settlement in the countryside.* London: HMSO.

Housing and Local Government, Ministry of 1970. *Development Plans: a manual on form and content.* London: HMSO.

Houston, A. M. 1975. Agricultural expansion, import substitution and the balance of payments. *J. Agric. Econ.* **26**, 351–66.

Irving, B. and L. Hilgendorf 1975. *Tied cottages in British agriculture.* London: Tavistock Institute of Human Relations.

Jackson, V. J. 1968. *Population in the countryside: growth and stagnation in the Cotswolds.* London: Frank Cass.

Jacobs, C. A. J. 1978. The role of the local planning authority in upland areas of Britain. In *The future of upland Britain*, R. B. Tranter (ed.). Reading: Centre for Agricultural Strategy.

Joint Unit for Research into the Urban Environment 1974. *Land availability and the residential land conversion process.* Birmingham: University of Aston.

Joint Unit for Research into the Urban Environment 1977. *Planning and land availability.* Birmingham: University of Aston.

Jones, A. G. 1977. *Village into town.* Exeter: Devon County Council and Exeter University.

Jones, A. G. 1979. *Rural recovery: has it begun?* Exeter: Devon County Council and Exeter University.

Jones, D. K. C. 1973. Problems of sand and gravel extraction. *Town and Country Planning* **41**, 368–73.

Jones, P. 1977. The spread of Dutch elm disease. *Town and Country Planning* **45**, 482–5.

Jowell, J. 1977. Bargaining in development control. *J. Plann. Environ. Law*, 414–33.

Juniper, B. 1977. Crops + Oil = Food. *Countryman* Autumn, 47–53.

Lake District National Park Office 1977. *Landscape surveys.* Kendal: The Office (unpublished typescript).

Law, C. M. and A. M. Warnes 1975. Life begins at sixty: the increase in regional retirement migration. *Town and Country Planning* **43**, 531–4.

Leat, D. J. K. 1978. *The role of pressure groups in rural planning.* Unpublished PhD thesis, University of Exeter.

Leonard, P. and R. Cobham 1977. The farming landscapes of England and Wales: a changing scene. *Landscape Planning* **4**, 205–36.

Levin, P. H. 1979. Highway inquiries: a study in governmental responsiveness. *Publ. Admin.* **57**, 21–49.

Locke, G. M. 1976. *The place of forestry in Scotland.* Edinburgh: Forestry Commission.

Loughlin, M. 1978. Bargaining as a tool in development control. *J. Plann. Environ. Law*, 290–5.

Low, N. 1973. Farming and the inner green belt. *Town Planning Review* **44**, 103–16.

McCrone, G. 1969. *Regional policy in Britain.* London: George Allen & Unwin.

Mackintosh, J. P. 1970. The problems of agricultural politics. *J. Agric. Econ.* **21**, 23–35.

Mackney, D. 1974. Land Use Capability Classification in the United Kingdom. In *Land Capability Classification*, Ministry of Agriculture. Technical Bulletin 30. London: HMSO.

MAFF *(Ministry of Agriculture, Fisheries and Food) 1946. National farm survey of England and Wales 1941–3.* London: HMSO.

MAFF 1958. *Warwickshire: a study of the loss of agricultural land for urban development.* London: HMSO.

MAFF 1968. *A century of agricultural statistics 1866–1966.* London: HMSO.

MAFF 1972a. *Forestry policy.* London: HMSO.

MAFF 1972b. *The appearance of farm buildings in the landscape.* London: HMSO.

MAFF 1975. *Conversion of agricultural land in the Slough/Hillingdon area.* London: HMSO.

MAFF 1976a. *Report on agricultural marketing schemes.* London: HMSO.

MAFF 1976b. *Wildlife conservation in semi-natural habitats on farms.* London: HMSO.

MAFF 1977a. *Land classification maps.* Pinner: MAFF.

MAFF 1977b. *The changing structure of agriculture 1968–75.* London: HMSO.

Marsh, J. S. 1977. *UK agricultural policy within the European Community.* Reading: Centre for Agricultural Strategy.

Martin, W. H. and S. Mason 1976. Leisure 1980 and beyond. *Long Range Planning* **9**, 58–65.

Mellanby, K. 1975. *Can Britain feed itself?* London: Merlin Press.

Mellanby, K. 1978. Crops – Oil = Not enough food. *Countryman* Spring, 45–50.

Minay, C. L. W. 1977. A new rural development agency. *Town and Country Planning* **45**, 439–43.

Moseley, M. 1974. *Growth centres in spatial planning.* Oxford: Pergamon.

Moseley, M. 1979. *Accessibility: the rural challenge.* London: Methuen.

Napolitan, L. 1975. The statistical activities of the Ministry of Agriculture. *J. Agric. Econ.* **26**, 1–19.

National Economic and Development Office 1977. *Agriculture into the 1980s: finance.* London: The Office.

National Farmers Union 1973. *A survey of current attitudes of people living in cities towards farmers and farming.* London: The Union.

Nature Conservancy Council 1977a. *A nature conservation review: towards implementation: A consultative paper.* London: The Council.

Nature Conservancy Council 1977b. *Annual report 1976–77.* London: HMSO.

Newby, H., C. Bell, P. Saunders and D. Rose 1977. Farmers' attitude to conservation. *Countryside Recreation Review* **2**, 23–30.

Newby, H. 1979. *Green and pleasant land.* London: Hutchinson.

Newton, J. P. 1977. Farming and forestry in the hills and uplands: competition or partnership. In *The future of upland Britain*, R. Tranter (ed.), 113–35. Reading: Centre for Agricultural Strategy.

Norfolk County Planning Department 1977. *Farmland tree survey of Norfolk.* Norwich: The Department.

North, J. and D. Spooner 1978. On the coal mining frontier. *Town and Country Planning* **46**, 155–63.

Northfield, Lord 1977. The role of the development commission. *Town and Country Planning* **45**, 304–307.

Northfield, Lord 1978. Rescuing rural England. *Countryman* 36–41.

Office of Population Censuses and Surveys 1973. *General household survey.* London: HMSO.

Office of Population Censuses and Surveys 1976a. *Changes in family building plans.* London: HMSO.

O'Riordan, T. 1979. The Broads Authority. *Town and Country Planning* **48**, 49–51.

Orton, C. 1972. The development of stratified sampling methods for the agricultural census. *J. R. Statist. Soc.* **135**, 307–15.

Orwin, C. 1970. *The reclamation of Exmoor*. Newton Abbot: David and Charles.

Parry, M. 1976. The mapping of abandoned farmland: an exploratory survey in south-east Scotland. *Geog. J.* **142**, 101–10.

Parry, M. 1977. *Mapping moorland change: a framework for land-use decisions in the Peak District*. Bakewell: The Peak National Park Authority.

Pearlman, J. J. 1978. Opencast coal mining. *J. Environ. Plann. Law* 234–9.

Penistan, M. 1972. *A hedgerow tree survey in East England Conservancy*. Edinburgh: The Forestry Commission.

Perman, D. 1973. *Cublington, a blueprint for resistance*. London: Bodley Head.

Peterken, G. and P. Harding 1975. Woodland conservation in eastern England. *Biological Conservation* **8**, 279–98.

Peters, G. 1970. Land use studies in Great Britain. *J. Agric. Econ.* **21**, 171–213.

Pollard, P., M. Hooper and N. Moore 1974. *Hedges*. London: Collins.

Porchester, Lord 1977. *A study of Exmoor*. London: HMSO.

Potter, D. 1975. Development control information: a systematic approach. *The Planner* **61**, 109–10.

Probert, G. and C. Hamersley 1979. Countryside management in Gwent. *The Planner* **65**, 10–13.

Rackham, O. 1976. *Trees and woodland in the British landscape*. London: Dent.

Rettig, S. 1976. An investigation into the problems of urban fringe agriculture in a green belt situation. *Planning Outlook* **19**, 50–74.

Rogers, A. W. 1976. Rural housing. In *Rural planning problems*, G. Cherry (ed.). 85–124. London: Leonard Hill.

Rogers, A. (ed.) 1978. *Urban growth, farmland losses and planning*. Ashford: Rural Geography Study Group.

Rossi, H. 1977. *Shaw's guide to the Rent (Agriculture) Act*. London: Shaw.

Rowe, S., H. Thomas, and S. Frost 1977. Hedge survey on a Hertfordshire farm. *Landscape Res.* **2**, 12–13.

Royal Town Planning Institute 1978. *Draft circular on design guidance*. London: The Institute.

Rural Planning Services 1978. *Motorways and agriculture*. Oxford: RPS.

Rural Planning Services 1979. *The Oxford Green Belt*. Oxford: RPS.

Samuels, A. 1978. Give me planning permission to replace my country cottage. *J. Plann. Environ. Law,* 94–7.

Shaw, M. J. (ed.) 1979. *Rural deprivation and planning*. Norwich: Geo Books.

Shoard, M. 1974. *Comments on the strategic settlement pattern for the South West*. London: CPRE.

Small, D. 1979. Forest for all seasons. *Geog. Magazine* **L1**, 620–8.

Smith, C. 1978. *National Parks*. London: Fabian Society.

Smith, R. T. 1975. *Power on the land*. London: Agricultural Engineers Association.

Snowdonia National Park 1975. *Rural land use management*. Penrhyndendraeth: The Park.

South-West Economic Planning Council 1974. *A strategic settlement pattern for the South-West*. London: HMSO.

South-West Economic Planning Council 1975a. *Survey of second homes in the South-West*. London: HMSO.

South-West Economic Planning Council 1975b. *Retirement to the South-West*. London: HMSO.

Stamp, L. D. 1961. *Applied geography*. London: Pelican.

Stamp, L. D. 1962. *The land of Britain, its use and misuse*. London: Longman.

Standing Conference of Rural Community Councils 1978. *The decline of rural services*. London: the Standing Conference.

Stephenson, I. 1975. *The law relating to agriculture*. London: Saxon House.

Storey, K. 1978. Development control performance revised. *District Councils Review*, 271–2.

Swinbank, A. 1978. *The British interest and the green pound*. Reading: Centre for Agricultural Strategy.

Tanner, M. F. 1976. Water resources and wetlands in England and Wales. In *Water planning and the regions*, P. Drudy (ed.). London: Regional Studies Association.

Thomas, D. 1964. Components of London's green belt. *J. Town Plann. Inst.* **50**, 434–9.

Thomson, J. M. 1978. The London Motorway Plan. In *Public participation in planning*, W. R. D. Sewell and J. T. Coppock (eds). London: John Wiley.

Thorburn, A. 1971. *Planning villages*. London: Estates Gazette.

Town and Country Planning, Ministry of 1950. *Landscape areas special development order*. London: HMSO.

Tracy, M. A. 1976. Fifty years of agricultural policy. *J. Agric. Econ.* **27**, 331–48.

Tranter, R. B. (ed.) 1978. *The future of upland Britain*. Reading: Centre for Agricultural Strategy.

Tucker, L. 1978. Planning agreements. *J. Plann. Environ. Law*, 806–809.

Turrall-Clarke, R. 1976. The agricultural condition of occupancy. *J. Plann. Environ. Law*, 71–4.

Vane, R. 1977. The registration of commons. *J. Plann. Environ. Law*, 352–8.

Vickery, D. 1978. Development control: a new sense of purpose. *The Planner* **64**, 24–6.

Waller, R. 1976. Taxation and the future of the farming structure. *Ecologist* **6**, 174–8.

Walston, Lord 1978. Land ownership. *J. Agric. Econ.* **29**, 321–8.

Watkins, O. 1979. Directions of local authorities in development control. *J. Plann. Environ. Law*, 431–4.

Weinstein, E. T. A. 1975. The movement of owner-occupier households between regions. *Regional Studies* **9**, 137–45.

Westmacott, R. and T. Worthington 1974. *New agricultural landscapes*. Cheltenham: The Countryside Commission.

Whitby, M. C. and K. J. Thomson 1979. Against coordination. *Town and Country Planning* **48**, 52.

Whitlock, R. 1975. Conservation by design. *Town and Country Planning* **43**, 260–2.

Wibberley, G. P. 1959. *Agriculture and urban growth: a study of the competition for rural land*. London: Michael Joseph.

Wibberley, G. P. 1972. Conflicts in the countryside. *Town and Country Planning* **40**, 259–64.

Wilkinson, G. 1978. *Epitaph for the elm*. London: Hutchinson.

Winifrith, J. 1974. The advisory services of the Ministry of Agriculture. *Agric. Admin.* **1**, 1–11.

Woodruffe, B. 1976. *Rural settlement policy and plans*. Oxford: Oxford University Press.

Index